# SpringerBriefs in Philosophy

For further volumes:
http://www.springer.com/series/10082

Erik Weber · Jeroen Van Bouwel
Leen De Vreese

# Scientific Explanation

Springer

Erik Weber
Jeroen Van Bouwel
Leen De Vreese
Centre for Logic and Philosophy of Science
Ghent University
Ghent
Belgium

ISSN 2211-4548          ISSN 2211-4556   (electronic)
ISBN 978-94-007-6445-3      ISBN 978-94-007-6446-0   (eBook)
DOI 10.1007/978-94-007-6446-0
Springer Dordrecht Heidelberg New York London

Library of Congress Control Number: 2013934540

Printed on acid-free paper

Springer is part of Springer Science+Business Media (www.springer.com)

# Contents

# Introduction

When scientists investigate why things happen, they aim at giving explanations. This practice of scientists can be analyzed by philosophers of science. In fact, there is a long tradition—which started in the second half of the twentieth century—of philosophical analysis of scientific explanation. Our motivation for writing this book is double. On the one hand, we think that the way in which philosophers of science have studied scientific explanation during most of this period, is flawed. We think it is important that new generations of philosophers of science see what went wrong in the past, so that they can avoid similar problems. Second, through our joint work on explanations in the social and in the biomedical sciences, we have gradually developed an alternative approach. This positive alternative has never been fully spelled out and never been presented in a systematic way: it is scattered among many papers published in several philosophy of science journals. Moreover, in these papers the focus lies on implementing the alternative approach, rather than on presenting and defending it. The latter will be done in this book.

Given this motivation, our first aim is to clarify what philosophers of explanation were doing and to show what went wrong in the past. Chapters 1 and 2 are devoted to this task. The title of the first chapter is *Theories of Scientific Explanation.* In this chapter, we first summarize the ideas of Carl Hempel, the godfather of this subdomain of the philosophy of science. Then we present the problems that other philosophers have raised in connection with Hempel's theory of explanation. Subsequently, we clarify how most of the major research traditions in the field that have emerged after Hempel, can be seen as different reactions to these problems. Finally, we discuss two such traditions (Philip Kitcher's unification approach and Wesley Salmon's causal-mechanical model) in more detail. This chapter does not give a complete overview. We give enough material so that the reader can grasp what these key players were doing (what their aims were, how they argued, …). The elements presented in Chap. 1 enable the reader to understand our criticism in Chap. 2.

The second chapter is entitled *How to Study Scientific Explanation?* Building on what is said in Chap. 1, this chapter investigates the working method of three important philosophers of explanation: Carl Hempel, Philip Kitcher, and Wesley Salmon. We argue that they do three things: (i) construct an explication in the sense of Rudolf Carnap, which then is used as a tool to make (ii) descriptive and

(iii) normative claims about the explanatory practice of scientists. We show that convincing arguments for the normative and descriptive claims are missing and that this should not come as a surprise, given the bold nature of their claims. At the end of this chapter we propose our alternative which we call the "pragmatic approach to scientific explanation." We clarify briefly what this approach consists in. Among other things, it involves more modest descriptive and normative claims than the traditional ones.

Our second aim is to elaborate and defend our alternative approach. We do this in Chaps. 3 and 4. In the third chapter, which is entitled *A Toolbox for Describing and Evaluating Explanatory Practices*, we develop a philosophical toolbox which contains different formats for explanation-seeking questions, different formats for explanations, and also what we call clusters of evaluative questions. The formats are tools for describing explanatory practices, but also for evaluating them (you cannot evaluate without knowing what is going on). The clusters of evaluative questions are of course tools for evaluating explanatory practices. In the last chapter, entitled *Examples of Descriptions and Evaluations of Explanatory Practices*, we provide illustrations of most parts of the toolbox. This chapter has an argumentative function (it shows that the toolbox is useful and that the pragmatic approach is viable) but also a paradigmatic function: it shows how the approach works, so other philosophers can apply it to their area of interest, if they are convinced of its fruitfulness.

We hope that our book will be useful for various audiences. It can be used as a textbook for intermediate philosophy of science courses. However, the way the book is set up also makes it an excellent study and research guide for advanced (M.A. and Ph.D.) students that work on the topic of scientific explanation. Finally, it is a handy source and reference book for senior researchers in the field of scientific explanations and—more generally—for all philosophers of science.

# Chapter 1
# Theories of Scientific Explanation

## 1.1 Introduction

In this chapter we first summarize the ideas of Carl Hempel, the godfather of this subdomain of the philosophy of science (Sect. 1.2). Then we present the problems that other philosophers have raised in connection with Hempel's theory of explanation (Sect. 1.3). Subsequently, we clarify how the major research traditions in the field that have emerged after Hempel, can be seen as different reactions to these problems (Sect. 1.4). Finally, we discuss two of these reactions in more detail. Philip Kitcher's unification account in Sect. 1.5 and Wesley Salmon's causal-mechanical account in Sect. 1.6.

This chapter gives the reader insights into some of the most important steps in the development of the domain till 1990. More recent developments (e.g. the work of James Woodward, Michael Strevens and the mechanistic approach) are treated in Chap. 3. But even for the period before 1990 we do not aim at completeness. For instance, the work of Bas van Fraassen is also discussed in Chap. 3, not here. The reason for this is that Chap. 2 plays a pivotal role in this book: we criticise the work of Hempel, Kitcher and Salmon at a meta-level and propose an alternative approach which is elaborated in Chaps. 3 and 4. Our criticism in Chap. 2 cannot be understood without knowledge of what these three philosophers have written on explanation. So we focus on their work. The work of other philosophers is saved for later: we try to integrate important insights into the toolbox we develop in Chap. 3.

## 1.2 Hempel's Models

### 1.2.1 The DN Model

We start with some definitions and terminology. According to Hempel, an explanation consists of an explanandum E (a description of the phenomenon to be explained) and an explanans (the statements that do the explaining). He distinguishes between true explanations and explanations that are well-confirmed (see e.g. Hempel 1965, p. 338). Both concepts are defined by means of the auxiliary concept of *potential explanans*. In order to get a grip on what a potential explanans is, Hempel developed two models: the deductive-nomological model (henceforth: DN model) and the inductive-statistical model (henceforth: IS model). In the DN model, a potential explanans is characterized as follows (cf. Hempel and Oppenheim 1948, part 3):

(DN) The ordered couple (L, C) constitutes a potential explanans for the singular sentence E if and only if

(1)  L is a purely universal sentence and C is a singular sentence,
(2)  E is deductively derivable from the conjunction L&C, and
(3)  E is not deductively derivable from C alone.

A purely universal sentence consists of one or more universal quantifiers, followed by an expression which contains no quantifiers and no individual constants. (L, C) is a true explanans for E if and only if (L, C) is a potential explanans for E and both L and C are true. (L, C) is a well-confirmed explanans for E if and only if (L, C) is a potential explanans for E and both L and C are well-confirmed.[1]

Let us consider an example. In volume I of *The Feynman Lectures on Physics* Chapter 26 deals with elementary optics. There we find a simple law about mirrors[2]:

> The simplest object is a mirror, and the law for a mirror is that when the light hits the mirror, it does not continue in a straight line, but bounces off the mirror into a new straight line[.] … The light striking a mirror travels in such a way that the two angles, between each beam and the mirror, are equal. (Feynman et al 2010, pp. 26–2)

Suppose that we have the following explanandum:

E     This reflected beam of light *a* has an angle of 45° relative to the mirror from which it bounced

According to definition (DN), the following construction is a *potential* explanans for this phenomenon:

C     The angle of incidence of *a* relative to the mirror was 45°
L     For all beams reflecting on mirrors: if the angle of incidence relative to the mirror is 45°, then the reflected beam also has an angle of 45° relative to the mirror

---

[1] An explanans may be true even if we don't have any evidence in favour of it. And a well-confirmed explanans may be false (which means that our evidence is incomplete and therefore misleading). For these reasons, Hempel introduces two concepts based on the concept of potential explanans.

[2] Note that Feynman assumes that the mirrors are flat.

If all these claims are true, we have a *true* explanans.

This example illustrates a general property of DN explanations. The first two conditions in (DN) imply that explanations have the form of a deductive argument. The simplest format is this:

L    $(\forall x)(Px \rightarrow Qx)$
C    $Pa$

_____

E    $Qa$

The mirror example fits this simple format.

In more complex explanations, C is a conjunction of atomic singular sentences $C_1, C_2, \ldots$ . The scheme is then:

L    $(\forall x)[(P_1x \wedge P_2x \wedge \ldots \wedge P_nx) \rightarrow Qx]$
$C_1$    $P_1a$
$C_2$    $P_2a$
...
$C_n$    $P_na$

_____

E    $Qa$

As an example we consider a case of thermal expansion. The expansion of aluminium rods is governed by the law $dL = 0.0000222 \times L_o \times dt$, where $dL$ is the expansion (in metre), $L_o$ the initial length (in metre) of the rod and $dt$ the temperature difference (in °C). 0.0000222 is the coefficient of linear thermal expansion of aluminium. With this background knowledge we can construct the following example:

$C_1$    This aluminium rod was heated from 50 to 250 °C
$C_2$    This aluminium rod has an initial length (at 50 °C) of 1 m
L    For all aluminium rods: if they are heated from 50 to 250 °C and their initial length is 1 m, then they are 4.44 mm longer at 250 °C

_____

E    This aluminium rod is 4.44 mm longer than it was before it was heated

The third condition in (DN) excludes circular arguments as explanations. The following deductive arguments are no explanations, though they satisfy the two first conditions:

L    $(\forall x)(Px \wedge Qx \rightarrow Rx)$
C    $Pa \wedge Qa$

_____

E    $Qa$

L    $(\forall x)(Px \rightarrow Px)$
C    $Pa$

_____

E    $Pa$

In both cases, E is derivable from C alone, so the argument is circular. Such circular arguments, though deductive, are not DN explanations.

## 1.2.2 The Value of Explanations

For Hempel, explanations are the instruments by which understanding of the world is achieved. So understanding the world is the intellectual benefit we expect to acquire by constructing explanations. What does this understanding of the world consist in? Hempel's answer is this:

> Thus a D-N Explanation answers the question 'Why did the explanandum-phenomenon occur?' by showing that the phenomenon resulted from certain particular circumstances, specified in $C_1, C_2,...,C_k$, in accordance with the laws $L_1, L_2,...,L_r$. By pointing this out, the argument shows that, given the particular circumstances and the laws in question, the occurrence of the phenomenon *was to be expected*; and it is in this sense that the explanation enables us to *understand why* the phenomenon occurred (Hempel 1965, p. 337; italics in original).

In other words: *understanding* must be identified with *expectability*, and expectability is the one and only intellectual benefit we can acquire by constructing explanations. This idea motivates the two first conditions in (DN).

## 1.2.3 The IS Model

Where the DN model is meant to capture the structure of deterministic explanations, the IS model intends to cover probabilistic explanations. Let us begin with a characterization given by Hempel:

> Explanations of particular facts or events by means of statistical laws thus present themselves as arguments that are inductive or probabilistic in the sense that the explanans confers upon the explanandum a more or less high degree of inductive support or of logical (inductive) probability; they will therefore be called inductive-statistical explanations; or I–S explanations (Hempel 1965, pp. 385–386).

Explanation is still linked with expectability, but in this case expectability comes in degrees. The idea of lawlike sentences thus has to be extended to account for *statistical laws* that have the conditional form $Prob(G|F) = r$, where $r$ denotes the probability that an object of the set F is also a member of the set G. The set F is called the *reference class* of this statistical law.

In its simplest form, an IS explanation is an argument with the following structure, analogous to DN explanations:

(**IS**)   L    $Prob(G|F) = r$

      C    Fb

      =============== [r]

      E    Gb

The notation is borrowed from Hempel: the double line indicates that the argument is inductive rather than deductive, and "[r]" represents the degree of inductive support that is conferred upon the conclusion by the premises. This argument explains the fact that object $b$ has property G by showing that this could be expected with probability $r$, given the fact that the statistical law $L$ holds, and that $b$ has property F. (Of course this structure can be extended to a more general schema in which the reference class of the conditional probability is determined by a conjunction of properties $F_1$ & $F_2$ & ...& $F_n$, and in which $b$ has the properties $F_1$, ..., $F_n$.).

Not all arguments of this form are IS explanations. As in the case of DN explanations we have to rule out the possibility of circular explanations. This can be done in exactly the same way: we have to require that E does not follow deductively from C. For instance, the following argument is not an IS explanation because it is circular:

L    $Prob(G|G) = 1$
C    Gb
       ============== [1]
E    Gb

Another extra condition is that $r > \varepsilon$, with $\varepsilon$ a chosen minimal degree of inductive support. Hempel calls this the *high probability requirement* (HPR). He requires that $r$ is high without specifying exactly how high. However, $r$ must always be higher than 0.5, otherwise we have an argument that makes us expect that the explanandum would not happen. In other words: the lower limit on our choice of $\varepsilon$ is 0.5.

In order to illustrate scheme IS and the HPR, we look at two examples. The following argument is an IS explanation provided that we set the value of $\varepsilon$ at between 0.5 and 0.8.

L    81 % of the 12–18 year old inhabitants of Flanders has a smartphone[3]
C    Jan is between 12 and 18 years old and lives in Flanders
       =============== [0,81]
E    Jan has a smartphone

The following construction (based on an example in Toulmin 1958, p. 111) is certainly not an IS explanation because HPR is not satisfied:

L    2 % of the inhabitants of Sweden are Roman Catholic
C    Petersen is a Swede
       =============== [0,02]
E    Petersen is a Roman Catholic

An example of an IS explanation given by Hempel is the explanation of why John Jones ($j$) recovered (R) from a streptococcus infection (S), when our

---

[3] Based on research done by Ghent University in the Spring of 2012.

knowledge system also contains the information that John was administered penicillin (P), and the probability ($r$, which is close enough to 1) of recovery from an infection given that penicillin is administered:

L    $Prob(R|S \& P) = r$
$C_1$    Sj
$C_2$    Pj
$==================$ [r]
E    Rj

Hempel, after introducing this example, immediately remarks that an important problem remains to be solved. Not all streptococcal infections can be cured by administering penicillin, and some streptococcus strains are even resistant to penicillin. The probability of recovery among the people who are treated with penicillin *and* are infected by a resistant strain is a number $s$ very close to 0 (equivalently, we could say that the probability of not recovering among these people is a number $1 - s$ which is very close to 1). If we would know that John Jones was infected by such a strain (Z), then we could give the following argument:

L    $Prob(\neg R|S \& P \& Z) = 1 - s$
$C_1$    Sj
$C_2$    Pj
$C_3$    Zj
$======================$ [1 – s]
E    ¬Rj

But now we are confronted with two strong inductive arguments, the premises of which could all be true at the same time, that give contradictory conclusions. This phenomenon is dubbed the *ambiguity of IS explanations* by Hempel.

The problem clearly has to do with the choice of the right reference class. Hempel's solution is the so-called *requirement of maximal specificity* (RMS) ('*b*', '*F*' and '*r*' as introduced above)[4]:

> [...] if $K$ is the set of all statements accepted at the given time, let $k$ be a sentence that is logically equivalent to $K$ [...]. Then, to be rationally acceptable in the knowledge situation represented by $K$, the proposed explanation [...] must meet the following condition (the requirement of maximal specificity): if [...] $k$ implies that $b$ belongs to a class $F_1$, and that $F_1$ is a subclass of F, then [...] $k$ must also imply a statement specifying the statistical probability of G in $F_1$, say
> $Prob(G| F_1) = r_1$.
> Here, $r_1$ must equal $r$ unless the probability statement just cited is simply a theorem of mathematical probability theory. (Hempel 1965, p. 400)

The unless-clause is needed to exclude the necessity of introducing $Prob(G|F\&G) = 1$ as the statistical law with the maximally specific reference

---

[4] We have adapted the notation in the quotation to ours—we also slightly altered Hempel's original condition, in which he allows for premises not contained in the knowledge system at the time of the explanation, since this doesn't make a difference to our discussion.

class. Hempel adds RMS as an extra condition: only inductive arguments that satisfy RMS can be IS explanations.

RMS implies that we always have to use all relevant information in IS explanations. To see how this works, we turn back to John Jones' recovery. If we know that $Zj$, the first explanation does not satisfy RMS: we know that John Jones not only belongs to class S&P but also to subclass S&P&Z; according to RMS an explanation that uses $Prob(R|S \& P) = r$ is rationally acceptable only if $Prob(\neg R|S \& P \& Z)$ also equals $r$; the latter is not the case. The second explanation does satisfy RMS (if there is nothing else we know about John Jones that changes the probabilities). So in this knowledge situation we have an IS explanation for $\neg Rj$, but not for $Rj$.

## 1.3  Problems for Hempel's Models

### 1.3.1  Accidental Generalisations

Hempel realised that he needed to distinguish between genuine laws and accidental generalizations. Consider the following two statements, apparently equivalent as far as their logical form goes (see Salmon 1989, p. 15):

(i)   No gold sphere has a mass greater than 100,000 kg.
(ii)  No enriched uranium sphere has a mass greater than 100,000 kg.

Whereas the second statement seems to be the expression of a lawful fact (the critical mass for enriched uranium is just a few kilograms; so we cannot create such a sphere because a much lighter sphere would already explode), the truth of the first statement seems to be a contingent matter of fact (it just happens to be the case that no one did produce such a sphere as yet; there is enough gold in the world and the sphere would not explode).

Hempel's (1965, p. 339) example is this: if we derive that Harry Smith is bald from the premises (i) that Harry Smith is a member of the Greenbury School Board for 1964, and (ii) that all members of the Greenbury School Board for 1964 are bald, this is not an explanation because we have used an accidental generalisation. Hempel's problem is that he has no viable account of how to distinguish between laws and accidental generalisation. He admits this, but does not take up the challenge.

### 1.3.2  Irrelevant Premises

Several people have offered counterexamples to Hempel's models.[5] The examples in the first group are inspired by the fact that the relation of logical deduction is monotonous: if you add premises to a deductive argument, the result is still a deductive

---

[5] Some nice footage illustrating these problems has been produced by the Centre for Logic and Philosophy of Science of Ghent University and is available at http://www.caeits2011.ugent.be/.

argument. This is not the case for explanations: most people will not regard arguments
with superfluous premises (that otherwise satisfy Hempel's conditions) as explanations.
Suppose we are convinced that the premises of the two following arguments are true:

L   $(\forall x)(Px \to Qx)$
C   Pa
_____

E   Qa

L   $(\forall x)(Px \land R \to Qx)$
C   Pa$\land$Ra

_____

E   Qa

Both are deductive arguments, but the second one contains a superfluous premise if we are convinced that $(\forall x)(Px \to Qx)$ is true.

A well-known example is:

This sample of table salt dissolves in water because I have hexed it, and all samples of
hexed salt dissolve in water.

The problem is that non-hexed salt also dissolves in water, so the premise about
hexing is superfluous. Another frequently used example is:

John Jones did not get pregnant during the last year because he took birth control pills,
and men who take birth control pills don't get pregnant.

Here the premise about the birth control pills is superfluous.

IS explanations also can contain superfluous premises, for instance:

It was almost certain that John Jones would recover from his cold in less than one week, because
he took vitamin C and almost all colds disappear within one week after taking vitamin C.

Here the problem is that almost all colds disappear within one week, even without vitamin C. In other words: the probability of recovery given that John Jones
takes vitamin C is not higher than the probability of recovery given that he does
not take vitamin C.

## 1.3.3 Asymmetry

Several people have argued that explanation is asymmetrical. Because arguments
can be reversed, this asymmetry is a problem for Hempel's models. Consider the
following questions and answers:

Question 1
Why does this flagpole have a shadow of 10 metres long?
Answer 1
The flagpole is 10 metres high. The sun is at 45° above the horizon.

Because light moves in a straight line, we can derive (by means of the Pythagorean
Theorem) that the flagpole has a shadow of 10 metres long.

Question 2
Why is this flagpole 10 metres high?

*Answer 2*
The flagpole has a shadow of 10 metres long. The sun is at 45° above the horizon.
Because light moves in a straight line, we can derive (by means of the Pythagorean Theorem) that the flagpole is 10 metres high.

The problem is that only the first argument is an intuitively acceptable explanation, while both answers are DN explanations in Hempel's sense.

Another famous example is the pendulum. The pendulum law ($P = 2\pi.\sqrt{L/g}$) describes the relation between the length of a pendulum (L) and its period (P). Consider the following questions and answers:

*Question 1*
Why does this pendulum have a period of 2.006 s?
*Answer 1*
The pendulum is 1 metre long. From this it can be derived by means of the pendulum law that its period is 2.006 s.

*Question 2*
Why does this pendulum have a length of 1 metre?
*Answer 2*
The pendulum has a period of 2.006 s. From this it can be derived by means of the pendulum law that it has a length of 1 metre.

Again we have two arguments satisfying Hempel's criteria, but only one intuitively acceptable explanation.

Before we go to the strategies for solving these problems, it must be noted that Hempel did not regard asymmetry as a problem: for him these examples show that we have bad intuitions about explanations. Once we realise that understanding equals expectability, we can get rid of the bad intuitions (because we realise that explanations are symmetrical).

# 1.4  Strategies for Solving the Problems

## 1.4.1  Causal Derivations

After describing the counterexamples mentioned in Sects. 1.3.2 and 1.3.3, Daniel Hausman writes:

> The most plausible diagnosis of these cases of DN arguments that are not explanations is that the premises in these arguments fail to focus on the *causes* of the phenomena described in their conclusions (Hausman 1998, p. 157).

Hausman's solution is straightforward: only derivations from causes (causal derivations) are explanatory, derivations from effects are not explanatory. The criterion for distinguishing them is independent alterability:

*Independent Alterability*
For every pair of variables, $X$ and $Y$, whose values are specified in a derivation, if the value of $X$ were changed by intervention, then the value of $Y$ would be unchanged (Hausman 1998, p. 167).

Derivations that satisfy this criterion are explanatory because they are causal. Note that Hausman holds on to the idea that all explanations are arguments. He simply adds a criterion to the DN model which ensures that the factors in the explanans are causes of the fact to be explained. This rules out irrelevant premises and guarantees asymmetry.

Hausman's criterion works as follows in the flagpole case. For X take the Sun's angle of elevation, and suppose we change it from 45° to, say, 20° (e.g. by way of mirrors). Then the Sun's angle of elevation is independently alterable relative to the other conditions specified in the derivation just in case it does not affect them. This holds for the flagpole's height, but not for its shadow. If we change the Sun's angle of elevation, the flagpole's height remains unaffected, but a change of the length of the shadow would automatically follow.

We do not explore this possibility further in this chapter. It will reappear in Chap. 3.

## 1.4.2 Positive Causal Factors

Nancy Cartwright (1983) shares Hausman's diagnosis. However, she considers his solution inadequate because she believes that explanations are not arguments. If explanations are not arguments, one cannot require that the derivation must be causal since there is no derivation. A typical example is this: one can explain why the mayor contracted paresis by invoking that he had untreated latent syphilis. Only 7 % of the people with latent untreated syphilis get paresis, so there is no inductive argument: the probability of the explanandum given the explanans is low. Nevertheless, this is a good explanation because syphilis is a positive causal factor: without latent syphilis, one cannot contract paresis, so the probability of paresis given latent syphilis is higher than the probability of paresis without latent syphilis.

Cartwright requires that the explanans contains causes and that it increases the probability of the explanandum:

> I consider eradicating the poison oak at the bottom of my garden by spraying it with defo-liant. The can of defoliant claims that it is 90 per cent effective; that is, the probability of a plant's dying given that it is sprayed is 0.9, and the probability of its surviving is 0.1. Here (…) only the probable outcome, and not the improbable, is explained by the spraying. One can explain why some plants died by remarking that they were sprayed with a power-ful defoliant; but this will not explain why some survive. (Cartwright 1983, p. 28)

So the causes that are included in the explanans are *positive* causal factors. *Negative* causal factors are to be avoided. This idea will not be explored further here. Like Hausman's causal derivations, it will come back in Chap. 3.

## 1.4.3 Positive and Negative Causal Factors

Paul Humphreys shares the diagnosis of Hausman and Cartwright. Like Cartrwight, he thinks that explanations are not arguments. However, he goes one step further: he considers the a posteriori probability of the explanandum[6] to be irrelevant. According to Humphreys, singular explanations have the following canonical form:

Y in S at t (occurred, was present) because of φ, despite ψ (Humphreys 1989a, p. 101).

Here φ is a (nonempty) list of terms referring to contributing causes of Y, ψ a (possibly empty) list of terms referring to counteracting causes of Y. Like Cartwright, Humphreys uses a probabilistic definition of causation. The difference is that, according to Humphreys (see 1989b, Sect. 1.5) the a posteriori probability of the explanandum is irrelevant for the quality of the explanation, so we should not impose criteria that make use of such values: no high probability values (Hempel, Hausman), no probability increase (Cartwright). And, in order to get a good explanation, we sometimes have to mention negative causal factors.

Humphreys gives an example in which he uses the following background knowledge:

The bubonic plague bacillus (*Yersinia pestis*) will, if left to develop unchecked in a human, produce death in between 50 and 90 % of cases. It is treatable with antibiotics such as tetracycline, which reduces the chance of mortality to between 5 and 10 % (Humphreys 1989a, p. 100).

Then he considers a person (Albert) who has died. About the form the explanation of this event should have, he writes:

[A]n appropriate response at the elementary level would be "Albert's death occurred because of his infection with the plague bacillus, despite the administration of tetracycline to him." (Humphreys 1989a, p. 100)

According to Humphreys it would be wrong to leave out the tetracycline, even though it made Albert's death less probable.

Like Hausman and Cartwright, Humphreys will not be discussed in the remainder of this chapter. We will return to his ideas in Chap. 3.

## 1.4.4 Unificationism

The fourth strategy is to develop a unificationist account of explanation. Philip Kitcher claims that besides the official position of Hempel on explanation and understanding, there is also an unofficial one:

What scientific explanation, especially theoretical explanation, aims at is not [an] intuitive and highly subjective kind of understanding, but an objective kind of insight that is achieved by a systematic unification, by exhibiting the phenomena as manifestations of common underlying structures and processes that conform to specific, testable basic principles (Hempel 1966, p. 83; quoted in Kitcher 1981, p. 508).

Kitcher ascribes to Hempel the view that, besides expectability, explanations can confer a second intellectual benefit upon us: unification. Whether or not this ascription is correct does not matter: the important thing is that Kitcher regards *unification* as the one and only benefit that explanations may produce and that he claims that he can solve

---

[6] This is the the probability of the explanandum given the causes mentioned in the explanans.

the problems mentioned here in Sects. 1.3.1–1.3.3. Kitcher's unification account is presented in detail in Sect. 1.5 of this chapter. As we will see, Kitcher preserves the idea that explanations are arguments and tries to solve the problems without invoking causation: his diagnosis of the problems is different.[7]

### 1.4.5  The Causal-Mechanical Model

According to Wesley Salmon, explaining…

> … involves the placing of the explanandum in a causal network consisting of relevant causal interactions that occurred previously and suitable causal processes that connect them to the fact-to-be-explained (Salmon 1984, p. 269).

Salmon's model is presented in detail in Sect. 1.6. We mainly focus there on his definitions of the crucial terms in this quote (causal interaction and causal processes) so that we can get a grip on what this so-called "causal-mechanical model" means.

Like Cartwright and Humphreys, Salmon gives up the idea that explanations are arguments. Instead, an explanation should describe the causal network that produced the explanandum. The elements of this causal network might lower the probability of the explanandum. So like Humphreys (and contrary to Cartwright) he rejects probability increase as a condition. The main difference between Salmon and Humphreys is the way they define causation: Humphreys uses a probabilistic definition (see Humphreys 1989a, p. 74), Salmon has developed a process theory of causation.

### 1.4.6  Overview

Schematically, the positions presented till now can be represented as follows (N/A means: not applicable):

|                                                        | Hempel | Hausman | Cartwright | Humphreys | Kitcher | Salmon |
| ------------------------------------------------------ | ------ | ------- | ---------- | --------- | ------- | ------ |
| Are explanations arguments?                            | Yes    | Yes     | No         | No        | Yes     | No     |
| Can explanations contain accidental generalisations?   | Yes    | No      | N/A        | N/A       | No      | N/A    |
| Can explanations contain irrelevant premises?          | Yes    | No      | N/A        | N/A       | No      | N/A    |
| Are explanations symmetrical?                          | Yes    | No      | No         | No        | No      | No     |
| Do explanations cite causes?                           | No     | Yes     | Yes        | Yes       | No      | Yes    |
| Do explanations increase the probability of the explanandum? | Yes | Yes | Yes        | No        | Yes     | No     |

---

[7] As a historical note, we should add that Kitcher's account builds on, amends and extends Michael Friedman's (1974) unificationist account.

# 1.5   Philip Kitcher's Unification Account of Explanation

## 1.5.1   The Idea of Unification

In Kitcher's view, unification is reached by constructing arguments in which parts of our knowledge are derived from other parts. An argument is "… a sequence of statements whose status (as a premise or as following from previous members in accordance with some specified rule) is clearly specified" (Kitcher 1989, p. 431). If K is a set of beliefs, Kitcher calls any set of arguments whose premises and conclusions belong to K, a *systematization of K*. Unification is reached by systematizing our set of beliefs, i.e. by constructing a systematization of it. However, unification comes in degrees: some systematizations of a given knowledge system K are better (provide more unification) than others. As soon as an individual has constructed a systematization of his set of beliefs, he or she has reached some minimal degree of unification. How high this degree is, depends on the quality of the systematization, which is to be judged by combining four criteria. To explain these criteria (what we will do in Sect. 1.5.3), we need the concept of *argument pattern* (see Sect. 1.5.2).

## 1.5.2   Argument Patterns

An argument pattern is a triple of (i) a sequence of schematic sentences, (ii) a set of sets of filling instructions, and (iii) a classification. A schematic sentence is an expression obtained by replacing some, but not necessarily all, of the nonlogical expressions in a sentence with dummy letters. The filling instructions are directions for replacing the dummy letters. An argument pattern contains one set of filling instructions for each entry of the sequence of schematic sentences. A classification describes the inferential characteristics of a sequence of schematic sentences. As an example, we consider the following argument pattern:

*Sequence of schematic sentences*:

(1) $a$ is a P.
(2) All P's are bald.
(3) $a$ is bald.

*Filling instructions*
(F1): In (1) $a$ must be replaced with the name of an individual, P with an arbitrary predicate.
(F2): In (2) $p$ must be replaced with the same predicate as in (1).
(F3): In (3) $a$ must be replaced with the same name of an individual as in (1).

*Classification*
(1) and (2) are premises, (3) follows from (1) and (2) by means of universal instantiation and modus ponens.

An example of an argument fitting this pattern is:

*Sequence of sentences*

(1) Harry Smith is a member of the Greenbury School Board.
(2) All members of the Greenbury School Board are bald.
(3) Harry Smith is bald.

*Classification*
(1) and (2) are premises, (3) follows from (1) and (2) by means of universal instantiation and modus ponens.

Note that an argument is a couple of a sequence of schematic sentences and a classification. Our example is also an instantiation of the following argument pattern:

*Sequence of schematic sentences*:

(1) $a$ is a P.
(2) All P's are Q.
(3) $a$ is Q.

*Filling instructions*

(F1): In (1) $a$ must be replaced with the name of an individual, P with an arbitrary predicate.
(F2): In (2) Q must be replaced with an arbitrary predicate; P with the same predicate as in (1).
(F3): In (3) $a$ must be replaced with the same name of an individual as in (1), Q with the same predicate as in (2).

*Classification*
(1) and (2) are premises, (3) follows from (1) and (2) by means of universal instantiation and modus ponens.

For each argument, we can find many patterns of which it is an instance. If a pattern and an argument are given, we can always decide whether the argument is an instantiation of the pattern: all we have to do is check whether the filling instructions have been appropriately executed.

## 1.5.3  Four Factors of Unifying Power

Kitcher uses argument patterns to distinguish explanations from non-explanatory arguments and to explicate what unifying power is. For an individual with knowledge K, an argument A can only be an explanation if it is acceptable relative to K (i.e. if the premises of A are members of K). But not all acceptable arguments are explanations: an acceptable argument is an explanation if and only if it instantiates an argument pattern that belongs to a privileged set of argument patterns. This set of argument patterns is privileged because it has a higher unifying power with respect to K than any other conceivable set of argument patterns. The unifying power of a set of argument patterns is determined by four factors.

Firstly, the unifying power of a set of argument patterns with respect to K varies directly with the number of accepted sentences (i.e. the number of elements of K) for which we can construct an acceptable deductive argument that instantiates a pattern in the set. So *ceteris paribus* the ideal set contains at least one argument pattern to account for each observed event. The second factor is paucity of patterns: the unifying power of a set of argument patterns varies conversely with the number of patterns in the set.

The two first factors may be illustrated by the following argument pattern (see Kitcher 1981, p. 528):

*Sequence of schematic sentences*

(1) God wants it to be the case that $\alpha$.
(2) What God wants to be the case is the case.
(3) It is the case that $\alpha$.

*Filling instructions*

(F1): $\alpha$ may be replaced with any accepted sentence.
(F3): $\alpha$ must be replaced with the same sentence as in (1).

*Classification*
(3) follows from the premises (1) and (2) by instantiation and modus ponens.

The set containing only this argument pattern scores very high with respect to the two factors: the number of patterns is extremely low, and everything can be explained with this pattern. Nevertheless, the unifying power of the set is zero because of the third factor: stringency of patterns. An argument pattern is more stringent than another if there are more similarities among the instantiations of the first than among instantiations of the latter. This means that (i) the first pattern contains less schematic letters or (ii) that the filling instructions of the first pattern set conditions on its instantiations which are more difficult to satisfy than those set by the filling instructions of the second pattern. The pattern above has no stringency, because there are no conditions to be met when instantiating the schematic letter $\alpha$. The first Harry Smith-pattern is more stringent than the second, because it does not contain the schematic letter Q. The unifying power of a set of argument patterns varies directly with the stringency of the patterns in the set.

Finally, the unifying power of a set of argument patterns also varies directly with the degree of similarity of its members. To clarify this last factor, we consider the following argument pattern:

*Sequence of schematic sentences*

(1) All members of species S that belong to category XX × XY, have phenotype $F_j$.
(2) *a* belongs to species S and to category XX × XY.
(3) *a* has phenotype $F_j$.

*Filling instructions*

(F1): S must be replaced with the name of a species, $F_j$ with a name of a phenotype characteristic for S, X and Y with names of genes characteristic for S.
(F2): a must be replaced with a name of an individual of S; the rest as in (1).
(F3): Same as in (1) and (2).

*Classification*
(3) follows from the premises (1) and (2) by instantiation and modus ponens.

XX × XY is a type of cross: an XX × XY-individual is a descendant from one parent with genotype XX and one parent with genotype XY. An example of an argument fitting this pattern is:

*Sequence of sentences*

(1) Humans which belong to category II × I$i$ do not suffer from the Tay-Sachs-disease.
(2) John is a human and belongs to category II × I$i$.
(3) John does not suffer from the Tay-Sachs-disease.

*Classification*
(3) follows from the premises (1) and (2) by instantiation and modus ponens.

The Tay-Sachs disease is also called "infantile amaurotic idiocy". In children suffering from this disease, an accumulation of complex lipids in the brains, milt and/or liver causes blindness (amaurosis) and idiocy immediately after birth; almost all children die within 10 years. The occurrence of the disease is determined by two genes, I and $i$: $ii$ individuals have the disease, II and I$i$ do not suffer from it. We now can explain what similarity among patterns means: in the argument pattern above we may substitute XX × XY for a different kind of cross, e.g. XX × XX or XX × YY. In this way we obtain argument patterns that are very similar to the original one but which will allow us to explain other events.

In Kitcher's view, unification is reached by constructing explanatory arguments, i.e. by constructing arguments in which parts of our knowledge are derived from other parts, and which instantiate a pattern of the privileged set. In his view, understanding amounts to showing that many events can be derived by using the same small set of stringent and similar argument patterns again and again. It is in this sense that scientific explanations unify our experiences.

## 1.5.4 Explanations Versus Non-Explanatory Arguments

How does Kitcher deal with the counterexamples to Hempel's models? Sets that do contain superfluous argument patterns are not optimal. Kitcher uses this criterion to rule out the use of accidental generalisations like 'All members of the Greenbury School Board for 1964 are bald.' It is possible to derive Harry Smith's baldness through a derivation which contains this premise. However, we also have to explain the baldness of people that are not member of the Greenbury School Board for 1964. The general biochemical argument pattern we set for that can also be used for Harry Smith and his fellow board members. So the argument pattern containing the accidental generalisation is superfluous, and its instantiations are arguments but no explanations. Irrelevant premises are excluded in a similar way: the pattern we use for non-hexed samples of salt can be used for hexed salt too.

What about asymmetry? The flagpole's height cannot be explained by the length of the flagpole's shadow because there is an alternative derivation of the flagpole's height which instantiates an argument pattern with greater unifying power. Just consider the fact that unlighted flagpoles also have heights. If we wanted to explain the height by the shadow, we would need a different explanation of the height in case it is dark. More generally, we would end up with different explanations for the height of lighted and unlighted things, and this is not the case for the derivation of the flagpole's height which invokes an origin and development argument pattern, rather than shadows. The same for the pendulum:

> Suppose now that we admit as explanatory a derivation of the length of a simple pendulum from a specification of the period. Then we shall have to explain the lengths of non-swinging bodies by employing quite a different style of explanation (an origin and development derivation). (Kitcher 1981, p. 525)

## 1.6 Wesley Salmon's Causal-Mechanical Model of Explanation

### 1.6.1 Etiological and Constitutive Explanations

Wesley Salmon distinguishes two kinds of explanations. First, there is *etiological* explanation, which

> … involves the placing of the explanandum in a causal network consisting of relevant causal interactions that occurred previously and suitable causal processes that connect them to the fact-to-be-explained. (Salmon 1984, p. 269)

As is clear from this quote, etiological explanations are explanations of facts. In Sects. 1.6.2 and 1.6.3 we clarify the two key concepts (causal interaction and causal process), and in Sect. 1.6.4 we give some examples of etiological explanations.

Salmon's second type is *constitutive* explanation. Constitutive explanations explain regularities rather than singular facts. And though they are causal, the way they explain is different from etiological explanations. Etiological explanations use causal *antecedents* (causal interactions and causal processes that occurred *before* the explanandum event took place; the word "previously" in the quote above is important). Constitutive explanations clarify how the regularity is constituted by an underlying causal mechanism. We give an example in Sect. 1.6.4.

### 1.6.2 Causal Interactions

The concept of *causal interaction* was introduced by Salmon in order to capture what he calls the innovative aspect of causation (as opposed to the conservative aspect, for which he developed the concept of *causal process*). We will adopt a definition that is very close to Salmon's original definition:

(CI)     At $t$ there is a causal interaction between objects $x$ and $y$ if and only if

   (1) there is an intersection between $x$ and $y$ at $t$ (i.e. they are in adjacent
       or identical spatial regions at $t$),
   (2) $x$ exhibits a characteristic P in an interval immediately before $t$, but a
       modified characteristic P' immediately after $t$,
   (3) $y$ exhibits a characteristic Q in an interval immediately before $t$, but a
       modified characteristic Q' immediately after $t$,
   (4) $x$ would have had P immediately after $t$ if the intersection would not
       have occurred, and
   (5) $y$ would have had Q immediately after $t$ if the intersection would not
       have occurred.

An object can be anything in the ontology of science (e.g. atoms, photons,...)
or common sense (humans, chairs, trees,...). This definition incorporates the
basic ideas of Salmon. The main difference is that, according to our definition,
interactions occur between two objects. In Salmon's definition, an interaction is
something that happens between two processes (see Salmon 1984, p. 171). This
modification was suggested in Dowe 1992. The modification is not substantial
(processes are world-lines of objects, i.e. collections of points on a space–time
diagram that represents the history of an object). The advantage of this terminology is that it is more convenient in analysing every-day and scientific causal talk.
   Because we stick close to Salmon's original definition, we can borrow his
examples. Collision is the prototype of causal interaction: the momentum of each
object is changed, this change would not have occurred without the collision, and
the new momentum is preserved in an interval immediately after the collision.
When a white light pulse goes through a piece of red glass, this intersection is
also a causal interaction: the light pulse becomes and remains red, while the filter
undergoes an increase in energy because it absorbs some of the light. The glass
retains some of the energy for some time beyond the actual moment of interaction. As an example of an intersection which is not a causal interaction, we consider two spots of light, one red and the other green, that are projected on a white
screen. The red spot moves diagonally across the screen from the lower left-hand
corner to the upper right-hand corner, while the green spot moves from the lower
right-hand corner to the upper left-hand corner. The spots meet momentarily at the
centre of the screen. At that moment, a yellow spot appears, but each spot resumes
its former colour as soon as it leaves the region of intersection. No modification of
colour persists beyond the intersection, so no causal interaction has occurred. One
might object to the last example that there are no objects involved (if one does not
regard light spots as objects) so the clauses (1)–(5) in the definitions are superfluous in this case. A clearer phenomenon that is not a causal interaction are two billiard balls lying next to each other (so condition (1) is satisfied).
   Let us now analyse how Salmon's concept of causal interaction can be used in
everyday or scientific causal talk. Suppose we want to make a claim about a causal
interaction, of the following form:

At $t$ there was a causal interaction between $x$ and $y$, in which $x$ acquired characteristic P' and lost characteristic P, and in which $y$ acquired characteristic Q' and lost characteristic Q.

Making claims about causal interaction presupposes a frame of reference that settles the level of description, the spatial scale and the timescale that will be used. The level of description determines the kinds of system we talk about (e.g. individuals or groups of individuals, macroscopic objects or elementary particles). The spatial scale determines the smallest unit of distance, and thus determines whether two systems are or are not in adjacent spatial regions (they are if the distance between them is smaller than the smallest unit of distance). Likewise, the timescale determines the smallest unit of time we will use, and thus allows us to distinguish between "sudden changes" as they occur in interactions, and slower evolution: we have a sudden change if and only if the change takes place in a period of time that is smaller than the smallest unit of time.

Let us clarify this by means of a series of examples. Consider a group of people in a seminar room. There is a speaker that tells his audience things that are really new to them. The seminar lasts 59 min. Now take the following frame of reference:

Objects = common sense macro objects
Space = rooms and multiples of them (floors, buildings)
Time = 1 h and multiples (days, weeks,....)

In this frame of reference, a set of interactions has occurred: the speaker and each member of his audience were in adjacent spatial regions (because they were in the same room), and a sudden change has occurred (they learned something new within 1 h).

Contrast this with a different frame of reference:

Objects = common sense macro objects
Space = 1 mm distance and multiples
Time = 5 s and multiples

In this frame of reference, the seminar does not constitute a causal interaction because the distances are too big and the changes are too slow. However, someone inoculating me to protect me against some disease would be causally interacting with me: there is less than 1 mm distance between my body and the needle of the syringe, and there is a sudden change in my body (within 5 s, it contains a fluid it did not contain before the interaction).

If we modify the last clause into:

Time = 0.5 s and multiples

the inoculation is not a causal interaction any more (because the change is too slow).

In this modified frame of reference, collisions between two billiard balls still constitute causal interactions. However, if we take smaller units of space and time, these collisions cease to be causal interactions.

We can draw two conclusions from these examples:

(1) Salmon's concept of causal interaction is a "skeleton concept": it cannot be applied to empirical phenomena until we supplement it with a frame of reference as outlined above.
(2) If something is a causal interaction given a frame of reference, refining the frame of reference is sufficient to ensure that the phenomenon fails to satisfy the conditions.

These characteristics of the use of the concept of causal interaction are a consequence of the vagueness of certain words in the definition. Salmon's vagueness has a great advantage: they entail that Salmon's definition is a polyvalent one that can be applied in many areas of science. Salmon himself expresses the hope that his theory is adequate for all scientific disciplines—including the physical, biological and social sciences—except quantum mechanics (see Salmon 1984, p. 278).

Before we turn to causal processes, it is useful to point out that scientists which use the concept of causal interaction can make a clever choice, or a choice that is not very clever. Consider a psychologist investigating the group of people in a seminar room mentioned above. The psychologist is interested in exchange of knowledge. If the spatial scale he chooses is too refined, none of the phenomena he is interested in will turn out to be causal interactions (because the people are too far away from each other). If the spatial scale is appropriate, some phenomena (e.g. successful exchanges in which one person learns something from another) come out as causal interactions, while other phenomena (e.g. failed communication) come out as intersections which are not causal interactions. Clever scientists in a given discipline will use an appropriate frame of reference for their domain: a frame of reference in which some phenomena in which they are interested constitute causal interactions, while others do not. However, the fact that not all frames of reference are equally good (and that most scientists quasi-automatically choose an appropriate frame of reference) should not let us forget that we cannot use the concept of causal interaction without choosing a frame of reference (see (1) above), and this choice has consequences for what we label as "causal interaction" and what not (see (2) above).

## 1.6.3 Causal Processes

Causal mechanisms are more than complexes of causal interactions. Causation also has a *conservative* aspect: properties acquired in causal interactions are often spontaneously preserved, in what Salmon calls *causal processes*. A process is a world line of an object (i.e. a collection of points on a space–time diagram representing the history of an object). Salmon divides processes into causal processes and pseudo-processes. Causal processes are capable of transmitting marks, pseudo-processes cannot transmit marks. Mark transmission is defined by Salmon as follows:

> Let P be a process that, in the absence of interactions with other processes, would remain uniform with respect to characteristic Q, which it would manifest consistently over an interval that includes both of the space–time points A and B (A ≠ B). Then a *mark* (consisting of a modification of Q into Q′), which has been introduced into process P by means of a single local interaction at point A, is *transmitted* to point B if P manifests the modification Q′ at B and at all stages of the process between A and B without additional interventions. (Salmon 1984, p. 148)

Salmon mentions material objects as examples of causal processes, but that is not accurate: processes (whether causal or not) are always *world lines of* objects. The object itself is not a process. Some objects have the capacity to transmit certain modifications of their structure to other spatiotemporal regions (like e.g. material objects). World lines of such objects are causal processes. The movement of a material object is a process (a world line of an object). Moreover, it is a causal process: the underlying object has a capacity to transmit marks.

An example which clearly illustrates the difference between causal processes and pseudo-processes is this:

> Consider a car traveling along a road on a sunny day. As the car moves at 100 km/hr, its shadow moves along the shoulder at the same speed. The moving car, like any material object, constitutes a causal process; the shadow is a pseudo-process. If the car collides with a stone wall, it will carry the marks of that collision – the dents and scratches – along with it long after the collision has taken place. If, however, only the shadow of the car collides with the stone wall, it will be deformed momentarily, but it will resume its normal shape just as soon as it has passed beyond the wall. Indeed, if the car passes a tall building that cuts it off from the sunlight, the shadow will be obliterated, but will pop up right back into existence as soon as the car returned to the direct sunlight. If, however, the car is totally obliterated – say, by an atomic bomb blast – it will not pop back up into existence as soon as the blast has subsided. (Salmon 1984, pp. 143–144)

Like the concept of causal interaction, the concept of causal process presupposes an underlying reference frame which specifies the objects and the time and space scales. Take for instance a person that has no contact with anybody else for 2 weeks. This person *can be seen* as transmitting a mark (for instance: the beliefs he has) even if he eats, drinks, breathes and interacts in various other ways with his non-human biological environment. The beliefs *can be seen* as spontaneously preserved because no intervention of other human beings (e.g. through communication) is necessary to preserve them. The phrase we put in italics ("can be seen") is crucial because strictly speaking there is no mark transmission. The requirement that the mark is preserved "without additional interventions" (cf. the last sentence of the definition of mark transmission) is not satisfied: the interactions with the biological environment are necessary to preserve the belief (without the interactions, the person dies and the belief disappears). However, it is possible to classify causal interactions into groups, e.g. biological interactions and non-biological interactions. To see why such a distinction is useful, consider a person A with a good memory, and a person B with a very bad memory (B needs repetition of the message to be remembered every hour). The claim that the content of the message is transmitted without additional *non-biological* interactions is true for A and false for B. The claim that the content of the message is preserved without additional interactions *tout court* is false for both A and B (without breathing, both A and B die). A clever scientist who is interested in phenomena related to memory will therefore disregard biological interactions when applying the concept of causal process. The result will be that he says that in A there is a mark that is transmitted, while in B there is no mark transmission.

### 1.6.4  Examples of Causal-Mechanical Explanations

We start with two examples of etiological explanations, which are elaborations of brief examples which Salmon gives (Salmon 1984, p. 178–179).

Consider a window that is broken at time $t$. You ask me why it is the case. If I tell you (a) that it was hit by a baseball ball at $t'$ (causal interaction) and that it stayed in broken condition after that (causal process), I have described a causal network as required in the definition of etiological explanations. This network can be enlarged by going back further in time: I can refer to an interaction between a baseball bat and the ball at time $t''$ ($t'' < t'$) in which the ball acquired the momentum that it preserved till it hit the window. Going further back in time, I can refer to someone throwing the ball in the direction of the person holding the bat.

As a second example, consider someone who has been killed with a gunshot in the brain. I explain why he is dead now by means of a causal interaction (the impact of the bullet in the head which caused immediate death) and a causal process (dead people do not resurrect spontaneously). Like in the case of the window, I can go back further in time: the killer triggering the gun (causal interaction), the causal interaction between the bullet and the firing pin of the gun and the motion of the bullet (a causal process).

In his example of a constitutive explanation (Salmon 1984, pp. 268–269) Salmon considers the causal relation between air humidity and the distance required to get airplanes off of the ground at an airport. The causal law to be explained is:

Increased relative humidity causes an increase in take-off distance, other things (such as the temperature, the altitude of the airport, the type of plane and the load it is carrying) being equal.

The constitutive explanation which Salmon gives for this law is:

In the case of a propeller-driven airplane, both the lift imparted by the wings and the thrust imparted by the propeller are manifestations of the Bernoulli principle. According to this principle, the greater the velocity of a moving fluid or gas, the smaller is the pressure that it exerts in the direction perpendicular to its direction of flow. The magnitude of this effect varies with the density of the fluid; the denser the fluid, the greater will be the lift and thrust. Consequently, the denser the air, the more readily will an airplane become airborne. In order to provide the explanation we are seeking, we must therefore show why humid air is *less dense* than dry air. Avogadro's law, which is embedded in the kinetic-molecular theory of gases, will enable us to do the job.

According to Avogadro's law, for fixed values of pressure and temperature, a given volume of gas contains the same number of molecules, regardless of the type of gas it is. A cubic meter of oxygen contains the same number of molecules as a cubic meter of nitrogen; a cubic meter of dry air contains the same number as a cubic meter of moist air. The main difference between moist air and dry air is that moist air contains more molecules of water and fewer nitrogen and oxygen molecules. The molecular weight of nitrogen ($N_2$) is 28, that of oxygen ($O_2$) is 32, and that of water ($H_2O$) is 18. When oxygen or nitrogen molecules are replaced with water molecules in a given volume of air, the mass is decreased; consequently, the density is lessened and the efficiency of the airfoils is reduced. This explains why a greater takeoff distance is needed when the humidity is higher (Salmon 1984, pp. 268–269).

## 1.7 Summary and Preview

In this chapter we first presented the DN and IS models of Carl Hempel (Sect. 1.2). After an overview of the problems that other philosophers have raised in connection with Hempel's models (Sect. 1.3) we clarified how most of the major theories in the field can be seen as different reactions to these problems (Sect. 1.4). In Sects. 1.5 and 1.6 we discussed two such theories in detail: Philip Kitcher's unification account and Wesley Salmon's causal-mechanical account.

The next step we will take (Chap. 2) is to look at the aims which Hempel, Kitcher and Salmon had when they were studying scientific explanation. This analysis will allow us to present our views on why and how philosophers should be engaged with scientific explanations. In Chap. 3 this view is elaborated theoretically, while in Chap. 4 it is illustrated using examples from various scientific disciplines.

## References

Cartwright N (1983) How the laws of physics lie. Clarendon Press, Oxford

Dowe P (1992) Wesley Salmon's process theory of causality and the conserved quantity theory. Philos Sci 59:195–216

Feynman R, Leighton R, Sands M (2010) The Feynman lectures on physics (the new millennium edition). Addison-Wesley Publishing Company, Reading

Friedman M (1974) Explanation and scientific understanding. J Philos 71:5–19

Hausman DM (1998) Causal asymmetries. Cambridge University Press, Cambridge

Hempel C (1965) Aspects of scientific explanation. In: Hempel C (ed) Aspects of scientific explanation and other essays in the philosophy of science. Free Press, New York, pp 331–496

Hempel C (1966) Philosophy of natural science. Prentice Hall, Englewood Cliffs

Hempel C, Oppenheim P (1948) Studies in the logic of explanation. In: Hempel C (1965a), p 245-290 (Originally in Philos Sci 15:135–175)

Humphreys P (1989a) The chances of explanation. Princeton University Press, Princeton

Humphreys P (1989b) Scientific explanation: the causes, some of the causes, and nothing but the causes. In: Kitcher P, Salmon W (eds) Scientific explanation. University of Minnesota Press, Minneapolis, pp 282–306

Kitcher P (1981) Explanatory unification. Philos Sci 48:507–531

Kitcher P (1989) Explanatory unification and the causal structure of the world. In: Kitcher P, Salmon W (eds) Scientific explanation. University of Minnesota Press, Minneapolis, pp 410–505

Salmon W (1984) Scientific explanation and the causal structure of the world. Princeton
      University Press, Princeton
Salmon W (1989) Four decades of scientific explanation. In: Kitcher P, Salmon W (eds)
      Scientific explanation. University of Minnesota Press, Minneapolis, pp 3–219
Toulmin S (1958) The uses of argument. Cambridge University Press, Cambridge

# Chapter 2
# How to Study Scientific Explanation?

## 2.1 Introduction

This chapter investigates the working-method of three important philosophers of explanation discussed in Chap. 1: Carl Hempel, Philip Kitcher and Wesley Salmon. We argue that they do three things: (i) construct an explication in the sense of Rudolf Carnap, which then is used as a tool to make (ii) descriptive and (iii) normative claims about the explanatory practice of scientists. In Sect. 2.2—which has a preliminary character—we present Carnap's view on what the task of explication is, on the requirements it has to satisfy and on its function. In Sect. 2.3 we show that Carl Hempel develops an explication of the concept of explanation and makes descriptive and normative claims with it. We also show that he fails to provide convincing arguments for these claims. In Sects. 2.4 and 2.5 we show that Philip Kitcher and Wesley Salmon had a similar working-method in their philosophical analysis of scientific explanation and failed at the same stage as Hempel: the arguments for the descriptive and normative claims are missing.

We think it is the responsibility of current philosophers of explanation to go on where Hempel, Kitcher and Salmon failed. However, we should go on in a clever way. We call this clever way the "pragmatic approach to scientific explanation." We clarify what this approach consists in and defend it in Sect. 2.6.

## 2.2 Rudolf Carnap on Explication

Carnap devotes Chap. 1 of his *Logical Foundations of Probability* to the notion of explication. The following quote summarises his view:

> According to these considerations, the task of explication may be characterized as follows. If a concept is given as explicandum, the task consists in finding another concept as its explicatum which fulfils the following requirements to a sufficient degree.

E. Weber et al., *Scientific Explanation*, SpringerBriefs in Philosophy,
DOI: 10.1007/978-94-007-6446-0_2, © The Author(s) 2013

1.  the explicatum is to be *similar to the explicandum* in such a way that, in most cases
    in which the explicandum has so far been used, the explicatum can be used; however,
    close similarity is not required, and considerable differences are permitted.
2.  The characterization of the explicatum, that is, the rules of its use (for instance, in the
    form of a definition), is to be given in an *exact* form, so as to introduce the explicatum
    into a well-connected system of scientific concepts.
3.  The explicatum is to be a *fruitful* concept, that is, useful for the formulation of many
    universal statements (empirical laws in the case of a nonlogical concept, logical theo-
    rems in the case of a logical concept).
4.  The explicatum should be as *simple* as possible; this means as simple as the more
    important requirements (1), (2), (3) permit (1950, p. 7).

According to Carnap, scientists often make explications. One of his examples is
the replacement of the concept of Fish by the concept of Piscis:

> That the explicandum Fish has been replaced by the explicatum Piscis does not mean that
> the former term can always be replaced by the latter; because of the difference in meaning
> just mentioned, this is obviously not the case. The former concept has been succeeded by
> the latter in this sense: the former is no longer necessary in scientific talk; most of what
> previously was said with the former can now be said with the help of the latter (though
> often in a different form, not by simple replacement).
>
> ...
> ... [T]he concept Piscis promised to be much more fruitful than any concept more similar
> to Fish. A scientific concept is the more fruitful the more it can be brought into connection
> with other concepts on the basis of observed facts; in other words, the more it can be used
> for the formulation of laws. The zoölogists found that the animals to which the concept
> Fish applies, that is, those living in water, have by far not as many other properties in
> common as the animals which live in water, are cold-blooded vertebrates, and have gills
> throughout life. Hence, the concept Piscis defined by these latter properties allows more
> general statements than any concept defined so as to be more similar to Fish; and this is
> what makes the concept Piscis more fruitful. (1950, p. 6)

His second scientific example is Temperature as explicatum of Warmer:

> The concept Temperature may be regarded as an explicatum for the comparative concept
> Warmer. The first of the requirements for explicata discussed in §3, that of similarity or
> correspondence to the explicandum, means in the present case the following: The concept
> Temperature is to be such that, in most cases, if x is warmer than y (in the prescientific
> sense, based on the heat sensations of the skin), then the temperature of x is higher than
> that of y. (1950, p. 12).

In both cases, the explicatum allows us to formulate more empirical laws than
the explicandum, and that is why they are fruitful.

According to Carnap, mathematicians also provide explications. For instance,
the definitions of natural number as given by Russell and Whitehead in their
*Principia Mathematica* are—in his view—explicata for prescientific arithmetical
terms (numerals and operations). The fruitfulness of these explicata lies in the fact
that "Peano's axioms become provable theorems in logic" (1950, p. 17). Logic is
to be understood here as first order predicate calculus plus set theory. Here the
fruitfulness lies in more theorems rather than more empirical laws (as already indi-
cated in condition 3 in the first quote).

Finally, Carnap claims that also philosophers often make explications. His aim
in his 1950 book is to develop a good explication of the concepts of probability

and (degree of) confirmation. Before we turn to the use of the concept of explication by philosophers of explanation, it is important to note that the fruitfulness of philosophical explications may be of a different nature than the fruitfulness of scientific and mathematical ones. It is possible that the fruitfulness of philosophical explications lies in the formulation of empirical generalisations (like scientific explications) and/or logical theorems (like mathematical explications). But a serious option to consider is that their fruitfulness lies in that they allow us to offer clear guidelines for scientists (i.e. we formulate norms with them, not empirical generalisations or logical theorems).

## 2.3   Carl Hempel's Working-Method

### 2.3.1   The First Stage: Explication

According to the DN model and IS model, we explain particular facts by *subsuming* them under a law. The law then, so to speak, "covers" the fact to be explained. For this reason Hempel's views on explanation are known as the *covering law* model. It is clear that Hempel sees his covering law model as an explication in the sense of Carnap. At the end of the long essay 'Aspects of Scientific Explanation' he writes:

> This construal, which has been set forth in detail in the preceding sections, does not claim simply to be descriptive of the explanations actually offered in empirical science; for—to mention but one reason—there is no sufficiently clear generally accepted understanding as to what counts as a scientific explanation. The construal here set forth is, rather, in the nature of an *explication*, which is intended to replace a familiar but vague and ambiguous notion by a more precisely characterized and systematically fruitful and illuminating one. Actually, our explicatory analysis has not even led to a full definition of a precise "explicatum"–concept of scientific explanation; it purports only to make explicit some especially important aspects of such a concept.
>
> Like any other explication, the construal here put forward has to be justified by appropriate arguments. In our case, these have to show that the proposed construal does justice to such accounts as are generally agreed to be instances of scientific explanation, and that it affords a basis for a systematically fruitful logical and methodological analysis of the explanatory procedures used in empirical science. (1965, pp. 488–489)

If we compare this with Carnap's criteria, we notice the following:

(1) Two of Carnap's criteria are explicitly mentioned: *fruitfulness* and *exactness*.
(2) Hempel does not mention *similarity to the explicandum* explicitly. However, proposals should do justice to what are generally agreed as instances of explanation (cf. the last sentence of the quote). "Doing justice to" is vague, but we think it can safely be assumed to correspond to Carnap's similarity: no exact match, but not too much dissimilarity.
(3) *Simplicity* is not mentioned. Is it self-evident for Hempel that philosophers should be as simple as possible, or does Hempel think that simplicity is not a

cognitive value for philosophers? Given how he proceeds, we bet on the first answer.

Based on this comparison, it is reasonable to assume that Hempel's idea of explication is the same as Carnap's.

## 2.3.2 Descriptive and Normative Claims

Hempel has a clear view on the fruitfulness of philosophical models of explanation. This view is expressed in the following quote:

> As is made clear by our previous discussions, these models are not meant to describe how working scientists actually formulate their explanatory accounts. Their purpose is rather to indicate in reasonably precise terms the logical structure and the rationale of various ways in which empirical science answers explanation-seeking why-questions. (1965, p. 412)
> The construal here broadly summarized is not, of course, susceptible to strict "proof"; its soundness has to be judged by the light it can shed on the rationale and force of explanatory accounts offered in different branches of empirical science. (1965, p. 425)

Hempel suggests here that his models give us insight in the logical structure of explanation (What do they look like?) and the rationale of explanations (Why do scientists construct them? Why are they valuable?). How this second aspect works can be seen in the following quote:

> Thus a D-N Explanation answers the question '*Why* did the explanandum-phenomenon occur?' by showing that the phenomenon resulted from certain particular circumstances, specified in $C_1, C_2, ..., C_k$, in accordance with the laws $L_1, L_2, ..., L_r$. By pointing this out, the argument shows that, given the particular circumstances and the laws in question, the occurrence of the phenomenon *was to be expected*; and it is in this sense that the explanation enables us to *understand why* the phenomenon occurred. (1965, p. 337).

In other words: explanations provide understanding and the covering law models—according to Hempel—provide insight into what understanding really is (cf. Sect. 1.2.2).

In claiming that his models give insight in the logical structure of explanations and in connecting DN and IS explanation with expectability and understanding, Hempel puts forward the following descriptive hypothesis:

> All scientists who have understanding as an aim really seek DN or IS explanations, so that the phenomenon they want to understand becomes expectable.

This hypothesis cannot be formulated without the explicatum. In this way, the explicatum is fruitful. In order to avoid misunderstanding of the descriptive claim, it is important to stress that it is *not* a claim about what scientists or laymen call explanations. It is a claim about the intentions of scientists, about the kind of understanding they are after.

In the quote above Hempel not only mentions "rationale" but also "force". One of the insights we get from his models—in his view—is that non-causal explanations can be as good as causal explanations: they can have the same explanatory force. Therefore, there is no reason to demand that a good scientific explanation should be causal. Hempel states that it is unclear "what reason there would be denying the status of explanation to all accounts invoking occurrences that temporally succeed the event to be explained" (pp. 353–354). In other words: a DN or IS explanation is sufficient for providing understanding. They are also necessary for Hempel, as the following quote illustrates:

> I think that all adequate scientific explanations and their everyday counterparts claim or presuppose implicitly the deductive or inductive subsumability of whatever is to be explained under general laws or theoretical principles. In the explanation of an individual occurrence, those general nomic principles are required to connect the explanandum event with other particulars [.] (1965, pp. 424–425)

This view can be summarised in the following normative claim:

> All scientists that are engaged in trying to understand the world should construct DN or IS explanations (and not necessarily something more specific, such as DN explanations citing causes).

This guideline cannot be formulated without the concepts of DN and IS explanation. Again, this explicatum is fruitful: we can formulate a guideline that we cannot formulate without it.

### 2.3.3  Hempel's Failures

Let us now have a look at the problems. A first problem is that Hempel does not provide arguments for his descriptive claim. He does not provide a database with records of relevant opinions of a large representative sample of scientists from all disciplines all over the world (i.e. their opinions on what kind of understanding is important). However, that is what he should have done if he wanted to build a convincing argument for his empirical claim. If we compare what he does with what Carnap has in mind with "Piscis" and other scientific explicata, Hempel resembles a scientist who develops a nice explicatum concept, formulates an interesting hypothesis with it and then stops instead of trying to collect empirical evidence. Hempel stops where the real challenge starts: gathering evidence for his hypothesis by systematically investigating the opinions of a large representative sample of scientists (this could be done by interviewing them or analysing their writings).

Our criticism of Hempel is similar to the complaints of experimental philosophers about appeals to intuition in analytic philosophy: claims about what laymen would say about a particular case are part of the argument for a philosophical theory, but traditional philosophers do not actually ask people what they think. More generally traditional philosophers (which experimental philosophers call armchair philosophers) do not use the standard empirical methods of the behavioural and social sciences when they make descriptive claims about human attitudes, opinions, behaviour, etc. (see Knobe 2004 for a concise statement of what experimental philosophy is).

The second problem is that Hempel does not give arguments for his normative claim. Why, for instance, are causal explanations not interesting in a special sense (i.e. in a sense different from the general one in which the set of DN explanations, of which they are a subset, are interesting)? Again, he stops where the real challenge begins.

## 2.4  Philip Kitcher's Working-Method

### 2.4.1  Kitcher Versus Hempel

In his paper 'Explanatory Unification' Philip Kitcher writes:

> Why should we want an account of explanation? Two reasons present themselves. Firstly, we would like to understand and to evaluate the popular claim that the natural sciences do not merely pile up unrelated items of knowledge of more or less practical significance, but that they increase our understanding of the world. A theory of explanation should show us *how* scientific explanation advances our understanding…. Secondly, an account of explanation ought to enable us to comprehend and to arbitrate disputes in past and present science. Embryonic theories are often defended by appeal to their explanatory power. A theory of explanation should enable us to judge the adequacy of the defense. (1981, p. 508)

The first reason Kitcher mentions is similar to what Hempel calls "rationale". However, Kitcher claims that Hempel has it all wrong: understanding consists in unification, not in expectability. They have a common aim, but Kitcher thinks that Hempel has failed. Kitcher does not mention the concept of explication explicitly, but his model of explanation satisfies Carnap's criteria: precise definition, similarity and fruitfulness (Kitcher formulates descriptive and normative hypotheses with it, see below).

On the descriptive side, Kitcher puts forward two claims. The first is a negative one, a claim against Hempel:

> Scientists who have understanding as an aim often seek something more specific than DN explanations. So Hempel's descriptive hypothesis is false.

This claim can be found at several places in Kitcher's work. For instance, in commenting on Hempel's covering law model he writes:

> Many derivations which are intuitively nonexplanatory meet the conditions of the model. (1981, p. 508)

The derivations which Kitcher has in mind are all the counterexamples which several philosophers have launched against Hempel (cf. Chap. 1). In order to grasp Kitcher's critique of Hempel, it is important to remember that they are not interested in what scientists and other people call explanations. They are interested in what understanding is. According to Kitcher, if you would present a scientist with the two derivations of the flagpole example then this scientist would say that the derivation of the length of the shadow from the height of the flagpole provides understanding, while the other derivation does not. So the first derivation would be judged as scientifically more interesting. Kitcher is convinced that scientists would judge similarly in other cases (e.g. the pendulum) and concludes that expectability is not the kind of understanding scientists are after. According to Hempel, scientists would judge that both derivations are equally interesting (the derivation of the height from the flagpole provides as much understanding as the other derivation). Similarly for the other cases such as the pendulum. Hempel concludes that expectability is the kind of understanding that scientists are after.

The problem with Kitcher's line of reasoning is the same as with Hempel's: he speculates about what scientists would answer when presented with a set of derivations and a corresponding question. He did not actually interview a representative sample of scientists. The result is that the dispute cannot be settled: both Hempel and Kitcher claim that if one would consult scientists, that would result in evidence supporting their hypotheses. They cannot both be right.

## 2.4.2 Kitcher's Positive Descriptive Claim

Kitcher also puts forward a positive descriptive claim:

> All scientists who have understanding as an aim really seek K explanations, so that the phenomena they want to understand become more unified.

"K explanation" stands for "Kitcher style explanation".[1] As we have seen in Chap. 1, an underlying idea of Kitcher is that, while all explanations are arguments, the converse is not true. And he uses argument patterns to distinguish explanations from non-explanatory arguments.

Here is a quote from Kitcher in which the positive descriptive claim is clearly present:

> Science advances our understanding of nature by showing us how to derive descriptions of many phenomena, using the same patterns of derivation again and again, and, in demonstrating this, it teaches us how to reduce the number of types of facts we have to accept as ultimate (or brute). (1989, p. 432; italics in original)

In Kitcher's work on explanation no evidence for this second, positive claim can be found. He has not surveyed a large sample of scientists of various disciplines, in order to investigate what their views on understanding are. He clarifies the meaning of his claim (by developing his account of unification) but—like Hempel—he stops where the real challenge starts: gathering empirical evidence for the positive descriptive hypothesis.

---

[1] There is no standard label, comparable to "DN" or "covering law" for Hempel's views, to denote explanations in the sense of Kitcher. So we introduce our own label here.

### 2.4.3  Kitcher's Normative Claim

Kitcher also puts forward a normative claim:

> All scientists that are engaged in trying to understand the world should construct K explanations (nothing more, nothing less).

This is a strong claim. On the one hand, it excludes the formats put forward by Cartwright, Humphreys and Salmon: a K explanation is always an argument, so what these non-argument views propose as explanation formats is "less" than a K explanations and thus not good enough. On the other hand, it implies that causal information is useless (causal explanations are "more" than K explanations, and we should not aim at this "more"). Here is a quote from Kitcher in which he clearly expresses a normative aim:

> The most general problem of scientific explanation is to determine the conditions which *must* be met if science is to be used in answering an explanation-seeking question *Q*. I shall restrict my attention to explanation-seeking why-questions, and I shall attempt to determine the conditions under which an argument whose conclusion is *S can be used* to answer the question "Why is it the case that *S*?" (1981, p. 510; emphasis added)

Kitcher does not argue for this normative claim either. Again, he stops where the real challenge starts.

## 2.5  Wesley Salmon's Working Method

In Chap. 1 of *Scientific Explanation and the Causal Structure of the World*, Wesley Salmon explicitly puts himself in the tradition of explication:

> Many philosophical studies, including the one to which this book is devoted, aim at providing reasonably precise explications of fundamental concepts[.] (1984, p. 4)

Like Hempel and Kitcher, he claims that his explication of the concept of scientific explanation provides insight into what understanding is:

> Our aim is to understand scientific understanding. We secure scientific understanding by providing scientific explanations; thus our main concern will be with the nature of scientific explanation. (1984, p. ix)
>
>   Scientific explanations can be given for such particular occurrences as the appearance of Halley's comet in 1759 or the crash of a DC-10 jet airliner in Chicago in 1979, as well as such general features of the world as the nearly elliptical orbits of planets or the electrical conductivity of copper. The chief aim of this book is to try to discover just *what scientific understanding of this sort consists in*. (1984, p. 3, emphasis added)

His ambitions are quite high. He hopes to have achieved this aim for all sciences except quantum mechanics:

> It is my hope that the causal theory of scientific explanation outlined in this book is reasonably adequate for the characterization of explanation in most scientific contexts—in the physical, biological, and social sciences—as long as we do not become involved in quantum mechanics. (1984, p. 278)

As we already mentioned in Chap. 1, according to Salmon, explaining…

> … involves the placing of the explanandum in a causal network consisting of relevant causal interactions that occurred previously and suitable causal processes that connect them to the fact-to-be-explained. (1984, p. 269)

The usual label for explanations of this type is "causal-mechanical explanation" (abbreviated here as CM explanation).

These quotes clearly reveal a descriptive aim, which is captured in the following hypothesis:

> All scientists (except maybe in QM) who have understanding as an aim really seek CM explanations.

The aim of Salmon's book is to clarify what CM-explanations are and to argue for this claim. Salmon only succeeds in the first task: he defines the crucial concepts, viz. causal interaction and causal process. Arguments for the descriptive claim are missing: there is only the hope to have succeeded (cf. the quote above, which comes from the very end of the book). Like Hempel and Kitcher, Salmon did not try to collect empirical evidence.

Of course we could treat Salmon charitably and claim that he should not have set himself descriptive aims that require large-scale empirical research for which philosophers do not have the (financial) resources and often not the methodological skills. He should have stayed where philosophers belong, in the critical and normative realm. In other words, he should have put forward the following claim:

> All scientists (except maybe in QM) who have understanding as an aim should seek CM explanations (nothing more, nothing less).

Salmon's book does not contain arguments for this claim, and so he does not do better in this respect than Hempel and Kitcher. The three philosophers put a lot of effort in clarifying what their (descriptive or normative) claims mean, but the arguments are missing.

## 2.6  A Pragmatic Approach to Studying Scientific Explanations

We think it is the responsibility of current philosophers of explanation to go on where Hempel, Kitcher and Salmon stopped. However, we should go on in a clever way. We call this clever way the "pragmatic approach to scientific explanation". Let us clarify what it consists in and defend it. A pragmatic approach to explanation has three guiding principles.

### 2.6.1  Context-Dependent Normative Claims

The first principle is: make context-dependent normative claims and argue for them. This means that you first look at how scientists actually explain in certain disciplines and within certain research traditions in a discipline and then reflect upon it. This reflection can have three forms. We can defend certain explanatory practices (e.g. when other scientists in the discipline deny the validity or usefulness of that practice) or criticise explanatory practices. And we can also try to provide guidelines for improving an explanatory practice. Examples of this can be found in Van Bouwel and Weber (2008a, b) and in De Vreese et al. (2010). These papers engage in debates on explanatory practices in the social sciences and in the biomedical sciences. Models of scientific explanation as developed by philosophers of science are used as tools to make context-dependent normative claims which reflect on (i.e. criticise, defend and/or try to improve) actual explanatory practices.

The fact that we use traditional models of explanation as toolbox is one of the reasons why we call our approach pragmatic (there are more reasons, see below). "Pragmatic" thus refers to an "instrumentalist" attitude towards models of explanation: we see them as tools in a toolbox, rather than as describing the "essence" of explanation or understanding. In doing this, we differ in one important respect from the normative endeavours of Hempel, Kitcher and Salmon: the claims we make are context-dependent, i.e. they relate to specific explanatory practices within specific disciplines. There are several reasons for this piecemeal approach. First, this approach remains close to the actual practice of scientists in the relevant disciplines and, therefore, its results will be easily accessible and applicable for those scientists (which increases the chance that philosophical reflection has an impact on scientific practice). Second, the only way to defend general context-independent normative claims is bottom-up (i.e. by generalising from context-dependent claims, if they turn out to be similar). Third (and related to the second reason), the context-dependent claims are more modest and therefore normally easier to argue for.

Another difference is that we not only put forward normative claims; we also argue for them in our papers. Chapter 4 includes material from the papers mentioned above and other papers. So the reader of this book will get a good idea of the kind of claims we make and the arguments we give.

## 2.6.2  Context-Dependent Descriptive Claims

The second principle is: try to make context-dependent descriptive claims and argue for them. On the one hand, this means that you try to describe the explanatory practice of scientists in a certain discipline or research tradition. For instance, Walmsley (2008) describes the practice of explanation of dynamical cognitive scientists and argues that they are constructing DN explanations (in Gervais and Weber (2011) it is nonetheless shown that this description is inaccurate). In these descriptions, traditional models of explanation are again used as tools in a toolbox (this is a second reason to call the approach pragmatic).

The advantage of this piecemeal approach (as compared to the general claims made by Hempel, Kitcher and Salmon) is that we do not need large samples of scientists we can interview or large samples of scientific writings we can analyse. In Chap. 4 we will use Richard Feynman's investigation of the Challenger disaster as an example. Our aim here is to find out what kind of explanations Feynman was giving in his report. That aim is very modest compared to the bold general claims that Hempel, Kitcher and Salmon make. The advantage is that we only have to look at Feynman's report, we do not have to care about other scientists in the world (nor about Feynman's other work). Another illustration is this. Suppose we want to describe how population geneticists explain micro-evolutions in populations. That is not a trivial task (for instance, we have to analyse all important textbooks on population genetics before we can make a claim that is relatively well supported) but at least we do not have to worry about what other biologists do or about what physicists, psychologists, sociologists etc. do. The general claims that Hempel, Kitcher and Salmon make require that we take into account all scientists in the world. That is logically possible, but not feasible at this moment. So at least for now we have to settle for the more modest descriptive endeavours we are proposing.

### 2.6.3 Epistemic Interests

The third principle is: take into account the epistemic interests (i.e. the reason scientists have for asking specific explanation-seeking questions) when trying to make context-dependent normative or descriptive claims about explanations. These epistemic interests have to be taken into account because they influence the type of explanation that is appropriate in a given context and also influence which properties of an explanation (e.g. depth, deductivity) are important and which not in the given context. The relation between epistemic interests, types of explanation and the value of properties of explanations is investigated in Weber et al. (2005) and Weber and Van Bouwel (2007). The use of epistemic interests constitutes a third reason to call our approach pragmatic: the idea of epistemic interests originates in the work of the American pragmatists (Charles Peirce, John Dewey) and is also clearly present in the work of the German pragmatist Jürgen Habermas (who uses the term "knowledge-constitutive interest", see Habermas 1978).

### 2.6.4 Methodological Neutrality Versus Methodological Commitment

Let us go back to Carnap's idea of explication. The pragmatic approach we defend still assumes that we develop explicata as a first step in the analysis of explanations: we propose to use them—in the same way as Hempel, Kitcher and Salmon did—to put forward descriptive and normative claims. The reason is that without explicata we have no precise claims. And if the meaning of the claims remains vague, we cannot develop decent arguments. So if we want to develop arguments, we need explicata. However, our approach requires that we give up an *implicit* assumption which Carnap makes. Carnap seems to assume that philosophers of science can get far by developing *one* explicatum for a given explicandum (e.g. probability or explanation). Hempel, Kitcher and Salmon certainly assume this: they are convinced that their explicatum will do all the work (maybe leaving out some "strange" fields like QM, cf. Salmon). Our pragmatic approach does not presuppose this monism and adopts pluralism as a heuristic guideline: in principle, every explicatum that has been developed by a philosopher of explanation in the past can be used to formulate a context-dependent descriptive or normative hypothesis that is then further investigated. In the end—after analysing all explanatory practices of scientists—it may turn out that all explicata except one are superfluous. However, an a priori choice for monism is an overhasty conclusion which is unfounded at this moment and thus—from a methodological point of view—a bad start for studying scientific explanation.

A different way of putting this is that our pragmatic approach is *methodologically neutral*: we use whatever is available (all the explicata that have been developed) as tools for analysing explanatory practices (both descriptively and normatively) and see what works best in a given context. Hempel, Kitcher and Salmon were *methodologically committed*: they first put a lot of effort in developing their models; and then of course they wanted to show that their models were good or better than the models of their rivals. So they were committed to showing that their efforts in developing an explicatum were not in vain. We do not have such an agenda, because we are eclectic with respect to existing models. Therefore nothing can distract us from our main aim: making context-dependent normative and descriptive claims and argue for them. This is an important advantage of our methodological neutrality.

## 2.6.5 *Pragmatic Approach Versus Pragmatic Theory*

Our pragmatic *approach to* explanations should not be confused with a pragmatic *theory of* explanation as developed by e.g. Bas van Fraassen (1980). According to a pragmatic (also known as erotetic) theory of explanation, an explanation is an answer to a why-question that uses the appropriate relevance relation. Which relevance relation is appropriate depends on the context. Though this theory is much vaguer than the model of e.g. Hempel, Kitcher and Salmon, it is still a *theory about what all explanations have in common*. Our proposal—the pragmatic *approach*—is situated at the meta-level: it is a proposal on *how to study* scientific explanation. One of the ingredients of the approach—as outlined above—is that we use models of explanations as instruments. So we stop making general claims about the nature of all scientific explanations. We do not even make vague claims like in the pragmatic theory of van Fraassen. We give up generality so that we can make precise, informative claims (which are valid in specific contexts). Van Fraassen goes in the opposite direction: he wants to retain generality and therefore ends up with vague claims.

## 2.7  Conclusion

In this chapter, we have presented the descriptive and normative claims on scientific explanation that Hempel, Kitcher and Salmon make by means of their explicata. We have identified a major shortcoming which they have in common: they do not gather empirical evidence for their descriptive hypotheses (this could be done by systematically investigating the writings of a large representative sample of scientists) and do not argue for their normative claims either.

Since it is—in our view—time for philosophers of explanation to improve their ways of studying scientific explanation, we presented our pragmatic approach to studying scientific explanation as a better way to proceed. The main new features of what we propose—compared to Hempel, Kitcher and Salmon—are: context-dependence of the normative and descriptive claims, providing arguments instead of merely putting forward hypotheses, and the role of epistemic interests.

## References

Carnap R (1950) Logical foundation of probability. Routledge and Keegan Paul, London

De Vreese L, Weber E, Van Bouwel J (2010) Explanatory pluralism in the medical sciences: theory and practice. Theor Med Bioeth 31:371–390

Gervais R, Weber E (2011) The covering law model applied to dynamical cognitive science: a comment on Joel Walmsley. Mind Mach 21:33–39

Habermas J (1978) Knowledge and Human Interests, 2nd edn (trans. of Erkenntnis und Interesse. Suhrkamp Verlag, Frankfurt 1968). Heinemann Educational Books, London

Hempel C (1965) Aspects of scientific explanation. In: Hempel C (ed) Aspects of scientific explanation and other essays in the philosophy of science. Free Press, New York, pp 331–496

Kitcher P (1981) Explanatory unification. Philos Sci 48:507–531

Kitcher P (1989) Explanatory unification and the causal structure of the World. In: Kitcher P, Salmon W (eds) Scientific explanation. University of Minnesota Press, Minneapolis, pp 410–505

Knobe J (2004) What is experimental philosophy? The Philosophers' Magazine, 28

Salmon W (1984) Scientific explanation and the causal structure of the World. Princeton University Press, Princeton

Van Bouwel J, Weber E (2008a) De-ontologizing the debate on social explanations: a pragmatic approach based on epistemic interests. Hum Stud 31:423–442

Van Bouwel J, Weber E (2008b) A pragmatic defence of non-relativistic explanatory pluralism in history and social science. Hist Theor 47:168–182

Van Fraassen B (1980) The scientific image. Oxford University Press, Oxford

Walmsley J (2008) Explanation in dynamical cognitive science. Mind Mach 18:331–348

Weber E, Van Bouwel J, Vanderbeeken R (2005) Forms of causal explanation. Found Sci 10:437–454

Weber E, Van Bouwel J (2007) Assessing the explanatory power of causal explanations. In: Persson J, Ylikoski P (eds) Rethinking explanation. Kluwer, Dordrecht, pp 109–118

# Chapter 3
# A Toolbox for Describing and Evaluating Explanatory Practices

## 3.1 Introduction

In this chapter we develop a toolbox for analysing explanatory practices. An analysis of an explanatory practice can be either a description or a description plus an evaluation (one cannot evaluate without knowing what is going on, so evaluation without description is impossible). Because explanations consist of an explanans and an explanandum, we need tools for analysing both parts. In Sect. 3.2 we introduce a set of important types of why-questions, ordered in four main categories. This section offers tools for describing the explananda that scientists are dealing with. In Sects. 3.3–3.6 we present possible formats for answers to explanation-seeking questions (each section is about one of the main categories distinguished in Sect. 3.2). That completes the toolbox we need for *describing* explanatory practices. Section 3.7 adds a normative component: tools for the *evaluation* of explanatory practices. These tools take the form of clusters of evaluative questions (we will present five such clusters).

Before we start with this, let us clarify how this chapter fits into the pragmatic approach we are advocating. Instead of general descriptive claims about all scientists, we propose to make descriptive claims about the explanatory practice of individual scientists or groups of scientists. Sections 3.2 till 3.6 contain the tools to do this. Section 3.7 contains the tools for making normative claims about the explanatory practices of scientists. In general, normative claims can have two forms: evaluative ("This act is good/bad") or prescriptive ("Do this!" Or "Don't do this!"). We offer a toolbox for making evaluative claims ("This explanatory practice is good/bad") but these can be translated into prescriptions ("Keep on doing this!" Or "Change your practice in this and this way!"). The scope of the claims we propose to make with the toolbox is restricted: they are about specific scientists in a specific context.

E. Weber et al., *Scientific Explanation*, SpringerBriefs in Philosophy,
DOI: 10.1007/978-94-007-6446-0_3, © The Author(s) 2013

## 3.2  Types of Explanation-Seeking Questions

### 3.2.1  Explanations of Particular Facts Versus Explanations of Regularities

Let us repeat a part of a quote from Salmon we already used in Chap. 2:

> Scientific explanations can be given for such occurrences as the appearance of Halley's comet in 1759 or the crash of a DC-10 jet airliner in Chicago in 1979, as well such general features of the world as the nearly elliptical orbits of planets or the electrical conductivity of copper. (1984, p. 3)

The distinction between questions about particular facts and questions about regularities is made by all philosophers of explanation, starting with Hempel. After giving some examples in which the DN-model is applied to particular facts, he writes:

> So far we considered only the explanation of particular events occurring at a certain time and place. But the question "Why?" may be raised also in regard to general laws. Thus … the question might be asked: Why does the propagation of light conform to the law of refraction? (Hempel and Oppenheim 1948, p. 247)

Many philosophers, after acknowledging its existence, disregard the second category completely and focus on the first. Other philosophers try to say something about both categories of explanations. In general, much more work has been done on explanations of facts than on explanations of regularities. In this chapter, we will discuss both categories.

### 3.2.2  Questions About Particular Facts

#### 3.2.2.1  Questions About Plain Facts

The most simple format of explanation-seeking questions about particular facts is this:

> Why is it the case that X?

Here X is a description of a particular fact. We use the label "questions about plain facts" to denote questions of this form. Examples are:

> Why did the French revolution occur in 1789?
> Why is Belgium a monarchy?
> Why did the space shuttle Challenger explode?

Hempel explicitly states that all explanation-seeking questions about facts have this format. After giving several examples, he writes:

> [A]nd in that case the explanatory problem can again be expressed in the form 'Why is it the case that p?', where the place of 'p' is occupied by an empirical statement specifying

the explanandum. Questions of this type will be called *explanation-seeking why-questions*. (1965b, p. 334)

As we have seen in Chap. 2, Kitcher's aim is …

… to determine the conditions under which an argument whose conclusion is $S$ can be used to answer the question "Why is it the case that $S$?". (1981, p. 510)

So he makes the same assumption as Hempel. Salmon too makes the same assumption, but less explicitly: all his examples fit this format and he consistently describes the explanandum as "the fact-to-be explained". (see e.g. Salmon 1984, p. 13 and pp. 15–19).

### 3.2.2.2  Contrastive Questions

Bas van Fraassen (1980) has challenged this assumption. According to van Fraassen, a why-question is always contrastive. The simplest form is "Why X rather than Y"? Such contrastive questions have two important features: the *topic* (in this case X) which is taken to be true and the *foil* (in this case Y) which is taken to be false. Here are some examples:

Why did John paint a portrait of the Queen, rather than a landscape?
Why did John rather than Bill paint a portrait of the Queen?

While van Fraassen introduced the term "topic", he used the term "contrast-class" instead of the now common term "foil". The contrast-class is a set of propositions that contains the (true) topic plus one or more false propositions. So he allows for more complex questions where the topic is contrasted with more than one alternative. For our purposes, the simple contrastive why-questions suffice.

Van Fraassen does not deny that people ask non-contrastive why-questions. For instance, it is possible that someone asks the following question:

Why did John paint a portrait of the Queen?

However, van Fraassen claims that such questions are inaccurate expressions of the cognitive problem the person has. The real problem is captured by a contrastive question, for instance one of the two questions above.

Van Fraassen's view entails a general and bold claim: *all* non-contrastive questions which scientists ask are inaccurate formulations of contrastive questions for which they really want an answer. This claim is very difficult to back up (it would require that we agree on who is a scientist and who not, and that we check all the questions ever asked by a scientist). However, we think that van Fraassen is right in two respects:

(1) Many why-questions that scientists ask are contrastive in nature.
(2) If a scientist asks a non-contrastive question, it is *sometimes* the case that this question does not adequately represent the cognitive problem of the scientist: the real problem he/she wants to tackle is contrastive.

Note the crucial shift from "all" to "sometimes" in claim (2). These two claims contain crucial insights we have to bear in mind when analysing explanatory practices in the sciences.

### 3.2.2.3 Resemblance Questions

Contrastive and non-contrastive why-questions both occur in scientific practice, so we include both types in our toolbox. However, we also include a third type of explanation-seeking questions: resemblance questions. These occur in scientific practice, but have been largely neglected in the philosophical analysis of scientific explanations. Elster (who is one of the few people that does take them into account) gives the following examples (2007, p. 12):

> Why do many people fail to claim social benefits they are entitled to?
> Why did nobody call the police in the Kitty Genovese case?[1]

An example we will discuss below in Sect. 3.5.3 is this:

> Why was there a revolution in Bourbon France, Manchu China and in Romanov Russia?

As is clear from these examples, resemblance questions focus on similarities between events, rather than on differences (as contrastive questions do) or on just one event (as questions about plain facts do).

## 3.2.3  Questions About Regularities

### 3.2.3.1 Examples and Types of Regularities

In order to get a feel of the subject-matter involved here, let us look at some examples and important groups of regularities (the groups are certainly not exhaustive). First, there are quantitative laws, typically expressed by means of a mathematical formula. An example is the Ideal Gas Law:

> For equal quantities of gas in a container, the product of pressure $P$ and volume $V$ is proportional to temperature $T$, with a proportionality constant $R$ (the ideal gas constant).

Second we have qualitative regularities describing sufficient condition relations between properties:

> Children of blue-eyed parents are always blue-eyed.
> All ravens are black.

---

[1] Elster clarifies the Kitty Genovese case as follows: "For more than half an hour on March 27, 1964, thirty-eight respectable, law-abiding citizens in Queens, New York, watched a killer stalk and stab a woman in three separate attacks in Kew Gardens. Twice their chatter and the sudden glow of their bedroom lights interrupted him and frightened him off. Each time he returned, sought her out, and stabbed her again. Not one person telephoned the police during the assault; one witness called after the woman was dead". (2007, p. 12, footnote 6)

Third, there are regularities in which capacities are ascribed to classes of objects, for instance:

Copper has the capacity to conduct electricity.

### 3.2.3.2 Types of Questions

Like with particular facts, we can distinguish three types of questions: plain questions, contrastive questions and resemblance questions. Examples of plain questions are:

Why do children of blue-eyed parents always have blue eyes?
Why are all ravens are black?

Examples of contrastive questions are:

Why do pigeons have the capacity to find their way back home, while other sedentary birds do not have that capacity?
Why do woodcocks migrate during the night, while pigeons cover long distances during the day?

Pigeon navigation will be used as an example in Sect. 4.5. Examples of resemblance questions are:

Why do both humans and desk calculators have the capacity to perform exact numerical calculations?
Why do bats and hedgehogs hibernate?

## 3.3 Possible Formats of Answers to Why-Questions About Plain Facts

### 3.3.1 Introduction

This section offers tools which can be used to analyse the attempts of scientists to answer why-questions about plain facts. We present five formats. Each format can be used to describe and evaluate the explanatory practice of scientists when they deal with questions of this type. We will provide brief illustrations of each format, but the proof of the pudding is in the eating: only by applying a tool from the toolbox in a systematic way to a scientific discipline or scientific episode (as will be done in Chap. 4) it can be shown that the tool is really useful. For instance, it is in principle possible that every time we try to use the format developed in Sect. 3.3.2, it turns out that what the scientists are doing does not fit the schedule (not even to a moderate degree). In that case, the format would be useless. Of course we do believe that every format we present here is useful somewhere. But we cannot prove that in this short book. For some formats, evidence for their usefulness will be given in Chap. 4, but in order to show that all formats are useful we need more examples than the length of this book allows.

## 3.3.2 The CDN Format

### 3.3.2.1 Structure

CDN stands for "causal deductive-nomological". This format is inspired by the ideas of Dan Hausman (see Sect. 1.4.1). The structure of CDN explanations is as follows:

(CDN)    $S_1$    Object $a$ has value $c_{1s}$ for variable $C_1$

       $S_2$    Object $a$ has value $c_{2s}$ for variable $C_2$

       ...

       $S_n$    Object $a$ has value $c_{ns}$ for variable $C_n$

       L    All objects in U with values $c_{1s}, c_{2s}, ..., c_{ns}$ for the corresponding variables have value $e_x$ for variable E

---

       Exp  Object $a$ (which belongs to population U) has value $e_x$ for variable E

We use "Exp" as label for the explanandum sentence to avoid confusion with the effect variable E (which is used in it). Note that the explanandum sentence also mentions a population. This is required because of an additional condition we impose: the variables $C_1$ till $C_n$ must be positive or negative causal factors for E in population U. Claims about causal relations between variables only make sense relative to a population[2], so this population must be mentioned in the explanandum. In sum, the CDN format is characterised by two features: the scheme (CDN) above plus the extra condition about causation.

For binary variables, positive and negative causal relevance can be defined as follows:

(PCF)    C is a positive causal factor for E in the population U if and only if $P_X(E)$ is greater than $P_K(E)$

(NCF)    C is a negative causal factor for E in the population U if and only if $P_X(E)$ is less than $P_K(E)$

These definitions are taken from Giere (1997) (Chap. 7). X is the hypothetical population which is obtained by changing, for every member of U that exhibits the value not-C, the value into C. K is the analogous hypothetical population in which all individuals that exhibit C are changed into not-C. $P_X(E)$ and $P_K(E)$ are the probability of E in respectively X and K.

---

[2] "Smoking causes lung cancer" is meaningless, while "Smoking causes lung cancer in humans" is a meaningful claim.

Let us give an example to clarify the definitions. If we claim that smoking (C) is a positive causal factor for lung cancer (E) in the Belgian population (U), this amounts to claiming that if every inhabitant of Belgium were forced to smoke there would be more lung cancers in Belgium than if everyone were forbidden to smoke. Conversely for the claim that smoking is a negative causal factor. (PCF) and (NCF) are not exhaustive: causal irrelevance is also a possibility. If we claim that smoking behaviour is causally irrelevant for the occurrence or absence of lung cancer this means that we believe that in the two hypothetical populations the incidence of lung cancer is equally high.

We choose Giere's definitions to implement the CDN model because they have certain technical merits and because they can easily be generalised to non-binary variables. We briefly comment on these points.

Technical merits relate to the way in which the definitions are formulated, rather than to the content of the definitions. The technical merits of Giere's definitions are the following:

(1) The relevant population is explicitly mentioned.
(2) The definition is "purely inferential", i.e. tells us what *follows from* causal beliefs without making any assumptions about how causal claims are confirmed (i.e. no assumptions about evidence for causal claims).
(3) The definition makes clear why causal knowledge is—in principle—practically useful (policy relevance, relevance for personal decisions, ...) because it defines causation in terms of what would happen in two *hypothetical* populations. Why should policy makers want causal knowledge? If we adopt Giere's definitions, the answer is clear: the hypothetical populations X and K correspond to populations a policy maker may create by means of some direct intervention (e.g. a ban on smoking, a mandatory inoculation...).

With respect to generalisation to non-binary variables, it is useful to point out that the so-called *potential outcome model*, which is well-known in the methodology of the social sciences, incorporates the same ideas as Giere's definitions (average effects and hypothetical populations) but is applicable to continuous variables. See e.g. Morgan and Winship (2007) (pp. 5–6 and pp. 31–37) for a presentation of this model.

### 3.3.2.2 Examples

Let us give some examples. Guinea pigs are black or white. The colour is determined by a pair of genes (B and *b*). With respect to phenotypic expression of these genes, we have three laws: BB and B*b* animals are black, *bb* animals are white. The following explanations fit the CDN format:

$S_1$  Animal *a* has inherited a B-gene from its father
$S_2$  Animal *a* has inherited a B-gene from its mother
L   All guinea pigs that inherit a B-gene from their father and from their mother are black

Exp Animal $a$ (which is a guinea pig) is black

| | |
|---|---|
| $S_1$ | Animal $b$ has inherited a B-gene from its father |
| $S_2$ | Animal $b$ has inherited a b-gene from its mother |
| L | All guinea pigs that inherit a B-gene from their father and a b-gene from their mother are black |

Exp Animal $b$ (which is a guinea pig) is black

Inheriting a B-gene is, according to the definition above, a positive causal factor for blackness (and negative for whiteness). Inheriting a b-gene is a positive causal factor for whiteness and negative for blackness.

Our last example is adapted from Hausman (1998, p. 166; we have adapted the notation, the content of the example is the same):

| | |
|---|---|
| $S_1$ | The height (H) is this flagpole equals $h^*$ |
| $S_2$ | The angle of elevation of the sun for this flagpole (A) equals $a^*$ |
| L | $H/S = \tan(A)$ |

Exp The length of the shadow of this flagpole (S) equals $s^*$ [where $s^* = h^*/\tan(a^*)$]

The variables used here are not binary, so here we would need a different definition of causal relevance between variables (e.g. an interventionist definition like in Woodward 2003).

### 3.3.3  The PCR Format

PCR stands for "positive causal relevance". This format is inspired by the ideas of Nancy Cartwright (see Sect. 1.4.2). We think that the following format is a useful tool for analysing a part of the explanatory practice of scientists:

| | |
|---|---|
| (PCR) $S_1$ | Object $a$ has value $c_{1s}$ for variable $C_1$ |
| $S_2$ | Object $a$ has value $c_{2s}$ for variable $C_2$ |
| ... | |
| $S_n$ | Object $a$ has value $c_{ns}$ for variable $C_n$ |
| P | $P(E = e_x \mid C_1 = c_{1s} \ \& \ C_2 = c_{2s} \ \& \ ... \ \& \ C_n = c_n)$ is higher than $P(E = e_x)$ |
| | ***** |
| Exp | Object $a$ (which belongs to population U) has value $e_x$ for variable E. |

We use the starred line (*****) here to separate explanans from explanandum. This type of line  indicates that what is above it explains what is below it. It does *not* indicate an inductive or deductive derivability relation (we have used the traditional notation ===== and ——— for those relations). Like with the CDN format, there is an extra condition: the variables $C_1$ till $C_n$ must be positive or negative causal factors for E in population U. There has to be at least one positive causal factor, otherwise P cannot be true. In general the positive causal factors must outweigh the negative causal factors if negative factors are present in the explanation (which is not always the case).

If we put Cartwright's poison oak example into this format, we get:

$S_1$    I have sprayed this tree in my garden with defoliant of brand X

P    Spraying with defoliant X increases the probability of dying in poison oaks
*****

Exp   This tree in my garden, which is a poison oak, is dead (i.e. has value "yes" for variable "dead")

This explanation mentions one positive causal factor and no negative ones. This is the simplest form a PCR explanation can have.

While Cartwright presents PCR as the only format which scientific explanations can have, for us it is only one tool in the toolbox, together with the CDN format and the other formats that still have to come.

### 3.3.4  The PNC Format

PNC stands for "positive and negative causal factors". This format is inspired by the ideas of Paul Humphreys (see Sect. 1.4.3). Like we did in Sects. 3.2 and 3.3 with Hausman and Cartwright, we will use Humphreys' ideas but drop all exclusivity claims about the resulting format: it is one of many tools in the toolbox.

The scheme which characterises the PNC format is this:

(PNC)    Pos-$S_1$   Object $a$ has value $c_{1s}$ for variable $C_1$.

Pos-$S_2$   Object $a$ has value $c_{2s}$ for variable $C_2$.

...

Pos-$S_n$   Object $a$ has value $c_{ns}$ for variable $C_n$.

Neg-$S_1$   Object $a$ has value $p_{1s}$ for variable $P_1$.

Neg-$S_2$   Object $a$ has value $p_{2s}$ for variable $P_2$.

...

Neg-$S_n$   Object $a$ has value $p_{ns}$ for variable $P_n$.
*****

Exp       Object $a$ (which belongs to population U) has value $e_x$ for variable E.

The variables $C_1$ till $C_n$ must be positive causal factors for E in population U, the variables $P_1$ till $P_n$ must be negative causal factors. If we put Humphreys' plague example into this format, we get:

Pos-$S_1$   Albert was infected with the plague bacillus

Neg-$S_1$   Albert took tetracycline
*****

Exp       Albert, who is a human, is dead (i.e. has value "yes" for variable "dead")

## 3.3.5 The Etiological Format

The etiological format is a more precise formulation of the ideas of Wesley Salmon as presented in Sect. 1.6. An etiological explanation is a chain of elementary etiological explanations. *Elementary etiological explanations* are defined as follows:

(EE)  An elementary etiological explanation of the fact that object $x$ at time $t$ has property P contains

(1) a description of a causal interaction, containing the claim that this interaction took place at $t'$ ($t' < t$) and the claim that $x$ has acquired P in this interaction, and
(2) the claim that P was preserved spontaneously in $x$ during the period $[t',t]$.

To clarify the concept of elementary etiological explanation, we reconsider the examples we already used in Sect. 1.6. Boys playing baseball have broken a window. An elementary etiological explanation of the fact that the window is broken at $t$, is:

At $t'$ ($t' < t$) there was a causal interaction between a ball and the window. In this interaction the window acquired the characteristic of being broken, and lost the characteristic of being intact. The ball loses its momentum and acquires a different momentum. The characteristic of being broken was spontaneously preserved by the window between $t'$ and $t$.

Suppose that someone is killed with a gun. An elementary etiological explanation of the fact that the victim is dead at $t$, is:

At $t'$ ($t' < t$) a bullet interacted with the body of the victim. In this interaction, the victim acquired the property of being dead and lost the property of being alive. The bullet loses its momentum and acquires a new, different momentum. The property of being dead is preserved spontaneously.

*Etiological explanations* are defined as follows:

(E)  A set of elementary etiological explanations is an etiological explanation of the fact that object $x$ at time $t$ has property P if and only if

(1) the set contains an elementary etiological explanation of the fact that object $x$ at time $t$ has property P, and
(2) each other element of the set is an elementary etiological explanation of an initial condition of a causal interaction described in another member of the set.

So etiological explanations are chains of elementary etiological explanations. In our first example, the momentum of the ball before the collision is one of the initial conditions of the interaction between the ball and the window. For this initial condition an elementary etiological explanation can be given: in its collision with the bat, the ball acquired a momentum which it preserved spontaneously till it hit the window. If we want to go on, we can describe the origin of the momentum of the ball before its interaction with the bat. In this way, we construct a chain of elementary etiological explanations which constitutes an etiological explanation of the fact that the window is broken. In the second example, the momentum of the bullet immediately before penetration, is an initial condition of the interaction. An elementary etiological explanation for this initial condition can be constructed by identifying the momentum of the bullet immediately before penetration with its momentum immediately after its causal interaction with the gun. This elementary etiological explanation is part of the chain that constitutes the etiological explanation.

### 3.3.6   Variations on Previously Described Formats

In Chap. 1 we saw that Hempel developed two models: the DN-model and the IS-model. In Sect. 3.3.2 of this chapter we adapted the DN-model by adding a causality requirement. A similar move is of course possible for the IS-model. Another interesting variation on formats described till now is what we call the causal default rule (CDR) format. It is a variation on the CDN-model in which the requirement of deductive derivability is dropped. However, instead of an inductive inference based on a (more or less) exact probability value (as we have it in IS-explanations), we have an inductive inference based on default rules, which are vaguer. Let us clarify this.

Default rules (e.g. "Birds usually fly") differ from universal generalizations (as they are used in DN-explanations) in that they allow exceptions (e.g.: Penguins don't fly). They also differ from probability statements (as they occur in IS-explanations) in that they do not specify the relative frequency of the exceptions and "normal" cases ("usually" can mean anything fairly close to probability 1). If we could specify the relative frequency, we could construe an IS explanation. However, the information required to do this is not always available. This is why it is useful to take into account a third possible form of covering law explanations: CDR explanations.

The CDR format is this:

(CDR)   $S_1$   Object $a$ has value $c_{1s}$ for variable $C_1$
      $S_2$   Object $a$ has value $c_{2s}$ for variable $C_2$
      …
      $S_n$   Object $a$ has value $c_{ns}$ for variable $C_n$
      DR   Objects in U with values $c_{1s}, c_{2s}, …, c_{ns}$ for the corresponding variables usually have value $e_x$ for variable E
      ====================== [usually]
      Exp   Object $a$ (which belongs to population U) has value $e_x$ for variable E

Examples will be given in Sect. 4.3.

## 3.4 Possible Formats of Answers to Contrastive Why-Questions

### 3.4.1 Introduction: Woodward's Desideratum

In his book *Making Things Happen*, Jim Woodward discusses a classic example (2003, p. 187):

All ravens are black.
*a* is a raven.

_____

*a* is black.

He claims that this is not a satisfactory explanation because it …

… doesn't tell us about the conditions under which raven *a* would be some other color than black. (2003, p. 193; italics in original)

One may debate whether this "counterfactual" desideratum is valid in all contexts (i.e. with respect to all types of explanation-seeking questions) but it is certainly plausible with respect to contrastive questions. Consider the question

Why did John steal a bicycle rather than a car?

An answer to this question is satisfactory only if it somehow tells us how the world should have been different (compared to the actual world) for the alternative to obtain. This is also the case for answers to the following question:

Why did John rather than Peter steal a bicycle?

In Sects. 3.4.2 till 3.4.4 we will present three formats for explanations of contrasts which satisfy this condition. In Sects. 3.4.2 and 3.4.3 we will deal with answers to questions of the form "Why X rather than Y?" in which the topic X is something that occurred and Y something that was possible but did not occur. The two questions above satisfy this condition. Another example is:

Why does John steal a bicycle rather than buy it?

In Sect. 3.4.4 we discuss answers to questions of the form "Why does *a* have property P while *b* has property P' ?". Examples are:

Why did John steal a bicycle while Peter stole a car?
Why did John steal a bicycle while Peter stole nothing?

In these examples, the question is about a contrast between two real facts.[3]

_____

[3] The questions of 4.2 and 4.3 have "rather than" as connective, the questions in 4.4 have "while". So it is easy to keep them apart.

### 3.4.2 From Reality to Alternative Scenarios

A first strategy to answer questions of the form "Why X rather than Y?" is this: first describe what really happened and then present alternative scenarios that would lead to the alternative result. To illustrate this strategy, we reuse the baseball example from Sect. 3.3.5. So we consider a group of boys who play baseball and break a window. A possible explanation of the window being broken at $t$ rather than intact, is:

> *Description*
> At $t'$ ($t' < t$) there was a causal interaction between a ball and the window. In this interaction the window acquired the characteristic of being broken, and lost the characteristic of being intact. The characteristic of being broken was spontaneously preserved by the window between $t'$ and $t$.
> *Alternative scenarios*
> In order to have an intact window at $t$, there should have been no interaction between the ball and the window, or an interaction with a different initial condition (lower speed of the ball, window less fragile) or an interaction in the period $[t', t]$ which overrules the window's capacity of spontaneously preserving the property of being broken (i.e.: the window is repaired or replaced).

This example fits a format we call *contrastive etiological* explanations (CE explanations). The general pattern of CE explanations is this:

> (CE)   *Description*
>          An etiological explanation (as defined in (EE)) of the fact that object $x$ at time $t$ has property P.
>          *Alternative scenarios*
>          In order for the contrastive fact to happen, the world should have been different in one of the aspects $A_1, \ldots, A_n$.

In CE explanations, the descriptive part has the form of an etiological explanation. It is possible to characterize similar, complementary contrastive formats, based on other types of descriptive parts (see Weber e.a. 2005 for examples). In Sect. 4.2 we give an elaborate example of a CE explanation.

### 3.4.3 From an Ideal Scenario to Reality

As the reader may expect, the second strategy follows the opposite direction: from a non-real alternative scenario to reality. More specifically, this strategy plays on the *discrepancies* between what really happened and the ideal pattern the explainer has in mind. These discrepancies can be spelled out by first determining the ideal pattern (this part of the explanation is called the *normative* part) and then point at the differences (this part is called the *differentiating* part). We look at an example.

Consider a game with the following set-up. An urn contains 20 red balls and 80 green ones. On the left of the urn is a green box, on the right is a red one. The games master draws a ball randomly and gives it to the blindfolded candidate who must put it in one of the boxes (so the candidate knows the colour and position of the target boxes, but does not know the colour of the ball). The rules of the game are:

If your ball is green and you put it in the green box, you receive €100.
If your ball is green and you put it in the red box, you receive €10.
If your ball is red and you put it in the green box, you receive €10.
If your ball is red and you put it in the red box, you receive €50.

The ideal pattern we have in mind could be:

*Normative part*
The candidate wants to win. He/she knows that the urn contains 20 red balls and 80 green ones and has correct causal beliefs (i.e., the candidate knows the rules of the game). The candidate knows that there are two options: put the ball in the red box on the right or in the green box on the left (i.e., the candidate has correct beliefs about available options). The candidate always maximizes expected utility (that is his/her decision method).

It is easy to calculate that a candidate satisfying this ideal pattern chooses the green box: the expected utility for the green box is €82 $[(0.8 \times 100) + (0.2 \times 10)]$ as opposed to €18 for the red box $[(0.80 \times 10) + (0.2 \times 50)]$. Now consider a candidate who chooses the red box. The following question arises:

Why did this candidate choose the red box, rather than the green one?

In order to explain this "abnormal" behaviour, we can point at differences between the candidate's real state of mind and the ideal pattern. This is what we do in the *differentiating part* of the explanations. For instance we may find out that the candidate did not know the rules, did not see the opportunities correctly, did not want to get as much money as possible and/or made a calculation error.

An elaborate example of this type of explanation will be given in Sect. 4.4.

### 3.4.4 Real Contrasts

If we try to apply Woodward's desideratum to questions about real contrasts, we encounter a problem: there is not one but two alternatives. Consider the question:

Why did John steal a bicycle while Peter stole nothing?

One alternative is that they both steal a car, the other is that they both do not steal anything. Before we can meaningfully ask whether an answer to this question satisfies Woodward's desideratum we have to decide which alternative we are interested in. As soon as we have done this, we can try to determine how the world should have been different (compared to the actual world) for the alternative to obtain. And if we are interested in both alternatives, we do this twice.

The format of these explanations can be clarified by means of the concepts introduced in Sect. 3.4.2. They have to contain two descriptive parts, comparable to what we required in (CE). Second, they must contain a *comparative* part which highlights the differences between the two descriptive parts. Finally, they must contain an alternative scenario which tells us how the world should have been different (compared to the actual world) for the alternative we are focussing on to obtain. In cases where we want two explanations because both alternatives are interesting, we have to construct two alternative scenarios.

## 3.5 Possible Formats of Answers to Resemblance Why-Questions

### 3.5.1 Introduction

Because resemblance questions are about similarities between events, explanations which aim at answering them must somehow *unify* these events. Unification is to be understood here in a very broad sense: somehow the explanation must tell us what the two (or more) events have in common. This idea can be spelled out in two different ways, which we call top-down and bottom-up unification respectively. Kitcher's conception of unification has been presented in Sect. 1.5. Skipper (1999) has argued that, besides this well-known way to elaborate the unification idea there is a second one:

> … I have provided the foundations of an alternative to Kitcher's way of understanding explanatory unification. Kitcher claims that unification is the reduction of types of facts scientists must accept in expressing their world view, and it proceeds through derivation of large numbers of statements about scientific phenomena from economies of argument schemata. I suggest that it is very much worth exploring whether unification can be conceived as the reduction of types of mechanism scientists must accept as targets of their theories and explanations, and whether it proceeds through the delineation of pervasive causal mechanisms via mechanism schemata. (1999, pp. S207–S208)

We will call Kitcher-style unification "top-down unification", and Skipper-style unification "bottom-up unification". We clarify these possibilities and the explanation formats that correspond to them in Sects. 3.5.2 and 3.5.3.[4]

### 3.5.2 Top-Down Unification

In top-down unification we show that the events to be unified are instances of the same (set of) law(s) of nature. In other words, *top-down unification* proceeds by subsuming different events under the same law(s). Unification of this type is achieved by constructing arguments (one for each event) in which it is shown that the events could be expected (cf. Hempel's DN explanations or—if one wants to

---

[4] These ideas were first presented in Weber et al. 2012.

avoid asymmetry problems—our CDN explanations) and in which the same laws are used. Let us look at an example. The explanation-seeking question is:

Why do Peter and Mary both have blood group A?

This question can be answered as follows:

*Unifying Law*
(L) All humans who belong to category $I^A I^A \times I^A I^O$ have blood group A.

*Application 1*
(L) All humans who belong to category $I^A I^A \times I^A I^O$ have blood group A.
($P_1$) Mary is a human and belongs to category $I^A I^A \times I^A I^O$.

---

($E_1$) Mary has blood group A.

*Application 2*
(L) All humans who belong to category $I^A I^A \times I^A I^O$ have blood group A.
($P_2$) Peter is a human and belongs to category $I^A I^A \times I^A I^O$.

---

($E_2$) Peter has blood group A.

This is the top-down way to answer the question: we show that *both* events could be expected on the basis of *one* law. Cases like this one (two events, one law) are the simplest instances of top-down unification. However, our definition allows more than two events and the use of more than one law (as long as *all* laws are used in *each* derivation). There is an important similarity between top-down unification as we conceive it and Kitcher's proposal: unification is reached by constructing deductive arguments, and these arguments are instances of more or less widely applicable argument patterns. Furthermore, the ideal is to construct as many arguments as possible (for as many explananda as possible) by means of a limited number of argument patterns. However, there are important differences. First, our definition requires that there is an actual *unification act* rather than the mere possibility to use the same pattern again and again. When you unify, you always unify something. Or, to be more precise, you unify at least two things. This aspect of the pragmatics of unification is not taken into account by Kitcher but is included in our definition.[5] Another difference is that we are inclined to require that the arguments have to fit the CDN format, which means that top-down unification would always be based on causal laws.[6]

---

[5] Halonen and Hintikka (1999) claim that unification is a criterion for theory choice but that it has nothing to do with explanation. They arrive at this wrong conclusion because they neglect unification acts. See Weber and Van Dyck (2002) for an argument against the position of Halonen and Hintikka.

[6] We cannot argue for this position here extensively. We refer the reader to Weber and Van Bouwel (2009) in which it is argued that it is very difficult to find intuitively acceptable non-causal explanations.

### 3.5.3   Bottom-Up Unification

*Bottom-up* unification consists in showing that the mechanisms which lead to different events contain similar or identical causal factors. This does not require subsumption under a law, so this kind of unification does *not* proceed by constructing arguments and showing that the events could be expected. We first give an example and then come back to the differences between bottom-up and top-down unification.

#### 3.5.3.1   Social Revolutions

The material we use is taken from an article of Michael Taylor on revolutionary collective action (Taylor 1988) which discusses Theda Skocpol's classic *States and Social Revolutions* (Skocpol 1979). By using comparative methods, Skocpol has formulated a so-called "structural" explanation for three successful modern social revolutions in agrarian-bureaucratic monarchies (the French, Russian and Chinese revolution). The structural conditions that, in her view, make a revolution possible (the revolutions can be successfully mounted only if these structural preconditions are met), relate to the incapacitation of the central state's machineries, especially the weakening of the state's repressive capacity. This weakening is caused by external military (and economic) pressure: because of the backward agrarian economy and the power of the landed upper class in the agrarian-bureaucratic monarchy, the attempt to increase the military power leads to a fiscal crisis. Escalating international competition and humiliations particularly symbolized by unexpected defeats in wars (which inspired autocratic authorities to attempt reforms) trigger social revolutions. The foreign military and economic pressure that triggered the respective social revolutions, were:

1. Bourbon *France* (1787–1789): financially exhausted after the War for American Independence and because of the competition with England in general.
2. Manchu *China* (1911–1916): the Sino-Japanese War (1895) and the Boxer debacle (1899–1901).
3. Romanov *Russia* (1917): massive defeats in World War I.

Skocpol's theory gives adequate answers to several contrastive questions, for instance:

Why did the French revolution start in 1789, rather than in 1750?

The answer is that the pressure was not big enough in 1750. However, we are interested here in (answers to) resemblance why-questions. A question of this type which Skocpol and Taylor address is:

Why was there a revolution in Bourbon France, Manchu China and in Romanov Russia?

If we ask this question, we want to know which causal factors these three revolutions have in common. Skocpol gives a part of the answer. She endorses the following principle:

External military/economic pressure is a necessary cause[7] of social revolutions.

According to Michael Taylor there is another causal factor which the three revolutions have in common, viz. a strong sense of community among the peasants:

When the peasant community was sufficiently strong, then, it provided a social basis for collective action, including revolutionary collective action and rebellions and other popular mobilizations. (1988, p. 68)

Taylor shows how the participation of vast numbers of peasants in collective action could be explained by means of economic incentives and selective social incentives. Without incentives to motivate participation, collective action is unlikely to occur even when large groups of people with common interests exist. Using this account of collective action, Taylor argues that peasant collective action in revolutions was based on community (as many historians have argued) and that this is mainly why the large numbers of people involved were able to overcome the free-rider problem familiar to students of collective action and opted for conditional cooperation.

Taylor's idea can be summarized in the following principle:

A strong sense of community is a necessary cause for social revolutions to occur.

This does not contradict Skocpol's principle: they are different but compatible claims about factors that occur in the causal ancestry of all social revolutions.

### 3.5.3.2 Discussion

Skocpol and Taylor refer to certain structural conditions present that made the revolutions possible. The common factors in the explanation do not entail DN-expectability of the outcome. It is perfectly possible that there are societies who are under pressure and have a strong rural community, without going into revolution. The common factors that provide the unification of the three mentioned revolutions are not sufficient for causing those revolutions. So we don't have top-down unification here. Instead we have bottom-up unification: the mechanisms leading to the events contain identical causal factors.

## 3.6  Possible Formats of Explanations of Regularities

### 3.6.1  Introduction

Explanations of regularities have been much less analysed by philosophers than explanations of particular facts. Unfortunately, we have to follow this bad habit

---

[7] A necessary cause is a positive causal factor as defined in Sect. 3.3.2 which satisfies an extra condition: $P_K(E) = 0$.

partially. We will only discuss answers to plain questions about regularities, leaving aside contrastive and comparative questions about regularities. We take this choice because, in order to discuss these types of questions, we would have to introduce a quite large conceptual apparatus that relates to function ascriptions (claims that ascribe a function to a component of a system). There is no room for that in this book.

In Sect. 3.6.2 we go back to Hempel and Oppenheim to find out what they had to say about explanations of regularities. We will see that they proposed something we can expect (subsumption under a more general law) but also saw a problem. In Sect. 3.6.3 we present a recent alternative for the covering law view: mechanistic explanations. In Sect. 3.6.4 we discuss explanation by aggregation, which is closely related but different from mechanistic explanation.

### 3.6.2  Covering Law Explanations of Regularities

As already mentioned in Chap. 2, the DN model and IS model entail that we explain particular facts by *subsuming* them under a law. The law then covers the fact to be explained and hence Hempel's views on explanation are known as the *covering law* model. Hempel does not limit the idea of a covering law to explanations of particular facts: in his view, regularities have to be explained by subsuming them under other laws (i.e. by showing that the explanandum regularity could have been expected given the law in the explanans). However, he did not elaborate this idea because he saw a problem he could not solve. Before we explain this problem, we give an example of what Hempel had in mind:

Covering law 1    All waves reflect.
Covering law 2    All sounds are waves.
_____

Expl.             All sounds reflect.

The problem which Hempel saw was circularity. In Chap. 1 we saw that, for Hempel, not all deductive arguments are explanations: circular arguments are excluded. He uses a simple method to avoid circularity in explanations of facts: if E is derivable from C alone, there is no explanation. In a famous footnote (1948, footnote 33) Hempel and Oppenheim note that there is a similar problem for explanations of regularities. Here is the (almost complete) footnote:

> The core of the difficulty can be indicated briefly by reference to an example: Kepler's laws, K, may be conjoined with Boyle's law, B, to a stronger law K•B; but derivation of K from the latter would not be considered as an explanation of the regularities stated in Kepler's laws; rather, it would be viewed as representing, in effect, a pointless "explanation" of Kepler's laws by themselves. The derivation of Kepler's laws from Newton's laws of motion and gravitation, on the other hand, would be recognised as a genuine explanation in terms of more comprehensive regularities, or so-called higher-level laws. The problem therefore arises of setting up clear-cut criteria for the distinction of levels of explanation or for a comparison of generalized sentences as to their comprehensiveness. The establishment of adequate criteria for this purpose is as yet an open problem. (1948, p. 273)

Note that for Hempel the circularity problem is a symptom of a more general problem: comprehensiveness. He has no solution for this general problem in 1948 and never provided one in his later work. Rather, like Strevens remarks (2008, pp. 219–220), he stopped worrying about it: in his 1965b, the problem is not mentioned any more. So the covering law model of explanations of regularities is presented as unproblematic in Hempel's later work.

Unlike Hempel, unificationists like Friedman (1974) and Philip Kitcher (1981) did tackle this problem. Kitcher uses stringency, which is one of the factors determining unificatory power, to do the job (1981, pp. 526–527). If we explain a law L by deriving it from its conjunction with Boyle's law, we use the following pattern:

$$\frac{B \wedge \alpha}{\alpha}$$

In this pattern B is Boyle's law and $\alpha$ an arbitrary law we want to explain. If we systematize our beliefs with this pattern, we do not need any other pattern. However, it is extremely non-stringent, like the God-pattern discussed in Sect. 1.5.3.

### 3.6.3  Mechanistic Explanations of Capacities

#### 3.6.3.1  Introduction

Though probably most philosophers of explanation realised that covering law explanations of regularities are difficult to find outside physics, an alternative was not developed until the end of the 20th century. This is how William Bechtel and Adele Abrahamsen describe the situation:

> The received view of scientific explanation in philosophy (the deductive-nomological or D-N model) holds that to explain a phenomenon is to subsume it under a law. However, most *actual* explanations in the life sciences do not appeal to laws specified in the D-N model. (2005, pp. 421–422)

The life sciences include biology (cell biology, genetics, …) but also e.g. neuroscience. Bechtel and Abrahamsen claim that the discrepancy is due to a focus on physics:

> Given the ubiquity of references to mechanism in biology, and sparseness of reference to laws, it is a curious fact that mechanistic explanation was mostly neglected in the literature of 20th century philosophy of science. This was due both to the emphasis placed on physics and to the way in which explanation in physics was construed. (2005, p. 423)

In this section we discuss the alternative that was developed from 2000 onwards: the mechanistic model of explanation.

#### 3.6.3.2  The Mechanistic Model

By a *capacity* we mean the ability of a system to exhibit some kind of behavior. The term ability is chosen for obvious reasons: it might be that for some reason a system does not exhibit this behavior. Thus, a failure of a system to perform a capacity at a specific time does not count against it having that capacity as such. Many regularities in biology, cognitive science, psychology, engineering sciences,... ascribe capacities to classes of systems. And one of the main tasks which scientists in these disciplines set themselves (besides establishing these regularities) is explaining them. So these scientists ask why-questions about regularities in which capacities are ascribed. Here are some examples:

Why do plants and bacteria have the capacity to convert carbon dioxide into organic compounds?
Why do humans have the capacity to see depth?

Starting with Machamer, Darden and Craver's seminal paper 'Thinking about Mechanisms' (2000), many philosophers have claimed that these capacities should be explained by means of a *mechanistic* explanation, which does not look like a covering law explanation at all. So mechanistic explanations are put forward as an alternative to covering law explanations. They are defined as follows:

A *mechanistic explanation* of a capacity is a description of the underlying mechanism.

We define mechanisms as follows:

A *mechanism* is a collection of entities and activities that are organized such that they realize the capacity.

This definition includes the three key terms which mechanists use: entities, activities and organization. A description of a mechanism is usually called a *model* of the mechanism. The core idea of mechanists is that, in order to have explanatory value, the model has to describe the mechanism in terms of its entities, its activities and the way these entities and activities are organized. Before we present an example, some characteristic quotes. Bechtel and Abrahamsen write:

A mechanism is a structure performing a function in virtue of its component parts, component operations, and their organization. The orchestrated functioning of the mechanism is responsible for one or more phenomena. (2005, p. 423)

Carl Craver writes:

[M]echanisms are entities and activities organized such that they exhibit the explanandum phenomenon. (2007, p. 6, italics removed).

These quotes show that mechanists have no unique way of defining what a mechanism is. However, there is a common core idea and our definition captures this core idea.

### 3.6.3.3 An Example

Our example is about an electrical circuit, which we label E:

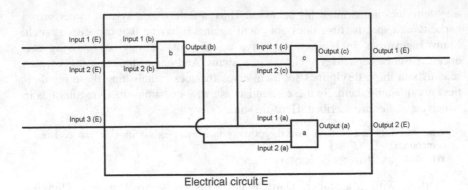

Electrical circuit E

Assume that everything inside the large rectangle is contained in an opaque box, so that only the three input wires and two output wires are visible. Assume also that we can somehow measure whether these wires are charged or not. Then an experiment can be performed to see whether there is a law connecting the states of the input wires with the states of the output wires. Suppose that such an experiment yields the following regularity:

If $input_1(E) = 1$, $input_2(E) = 0$ and $input_3(E) = 1$, then $output_1(E) = 0$ and $output_2(E) = 1$.

"$Input_1(E) = 1$" is shorthand for "The first input wire of E is charged", $input_2(E) = 0$ for "The second input wire of E is not charged", etc. This regularity ascribes a capacity (the capacity to produce a specific output given a specific input) to the system E.

In order to give a mechanistic explanation of this capacity, we have to open the box. If we do this, we discover that E contains three binary gates ($a$, $b$ and $c$) and several wires. This is what we find out about the *entities* of the mechanism. Each of the gates can be taken out of the circuit, so we can investigate their individual behaviour. Assume that such test gives the following results:

$a$ is an AND gate.
$b$ is an XOR gate.
$c$ is an XOR gate.

An AND gate has output 1 if and only if both inputs are 1. An XOR gate (exclusive OR) has output 1 if and only if the values of the inputs are different. These are claims about the *activities* of the entities. Finally, we can describe the *organisation* of the circuit:

The circuit is wired such that:
$output(b) = input_2(a)$.
$output(b) = input_1(c)$.
$input_1(E) = input_1(b)$.
$input_2(E) = input_2(b)$.
$input_3(E) = input_1(a) = input_2(c)$.
$output_1(E) = output(c)$.
$output_2(E) = output(a)$.

### 3.6.4  Explanation by Aggregation

Though it is very common in physics, explanation by aggregation is much less dis-cussed by philosophers of science than mechanistic explanations (Kuipers 2001, pp. 82–104 is an exception). Explanation by aggregation is similar to mechanis-tic explanation in that it invokes entities and activities at a lower level (i.e. there is decomposition of the system of which the behaviour is explained). However, it differs from mechanistic explanation in that it does not invoke organisation. An example is the explanation of Boyle's law by means of the kinetic theory of gases. We first give an outline of this explanation and then use it to clarify the character-istics of explanation by aggregation and its relation with mechanistic explanations.

Boyle's law states that for gases held at a fixed temperature and mass, the product of pressure (P) and volume (V) is constant ($P \times V = k$, where $k$ depends on the temperature and amount of gas). This law (and many other gas laws such as the more general ideal gas law mentioned in Sect. 3.2.3) can be explained by means of a kinetic model of gases (which fits into the general kinetic theory of matter; the overall theory also deals with properties of fluids and solids). Holton and Roller (1958) give a summary of the assumptions of the kinetic model of gases. Their summary is perfectly suited for our purposes. Here are the six assumptions they list:

(a) *Gases consist of molecules.* The first assumption, of course, is that a gas con-sists of molecules, each made up of an atom or a group of atoms [.]

…

(b) *The size of the molecules is negligible.* The actual volume occupied by the molecules is assumed to be negligible compared with the space between them.

…

(c) *The number of molecules is very large.* We may assume, in our model, that the number of molecules in any sample of gas is almost inconceivably large.

…

(d) *Molecules are in random motion.* Fundamental in the kinetic theory is the assumption that the molecules are in perpetual motion and traveling in so ran-dom a fashion that the number moving in any one direction is, on the average, the same as that moving in any other direction.

…

(e) *The forces between the molecules are negligible.* The molecules do not exert appreciable forces on one another except during collisions.

…

(f) *Collisions are perfectly elastic.* Returning now to our gas model, we assume that the collisions between molecules and with the walls of the container are perfectly elastic …; in other words, we assume that the kinetic energy of a molecule just after an impact is the same as immediately before the impact. (1958, pp. 433–436)

The gas model nicely illustrates the first characterising property of explanation by aggregation: like mechanistic explanation, it invokes entities and activities at a lower level. The entities involved here are the gas molecules and the walls of the container. The activities are: collisions and linear motion at constant velocity.

The second property—absence of organisation—can be clarified by the notion of function as it is used in biology. Or rather, by opposing two of the four function concepts which are defined in Wouters (2003). In that paper, Arno Wouters claims that biologists use four different notions of function. We need the first two:

1. function as activity (function$_1$)—what an organism, part, organ, or substance by itself does or is capable of doing.
2. function as biological role (function$_2$)—the way in which an item or activity contributes to a complex activity or capacity of an organism. (2003, p. 635)

The example he gives is this:

In the case of the heart of vertebrates, for example, the rhythmic contraction of the heart is an activity which the heart performs by itself (function$_1$). This activity enables the heart to perform its biological role (function$_2$) in the circulatory system, namely pumping the blood. (2003, p. 635)

If we replace "organism" with "system", Wouters' second notion can be used outside biology as well: activities of entities may have e.g. a physical, technical, psychological or social role. In the electrical circuit of Sect. 3.6.3.3, the gates have a technical function. For instance, the function of gate $b$ is to convert two system inputs (input$_1$(E) and input$_2$(E)) into an input for gate $a$ (input$_2$($a$)) and an input for gate $c$ (input$_1$($c$)). In aggregation explanations, the activities of entities do not have a function in this sense. If you take an arbitrary molecule in a gas, it is not possible to ascribe a specific function to it because all the molecules behave in the same way (cf. the assumptions (e) and (f) above). For instance, it is not correct to say that one molecule has the function of bouncing against the upper side of the container, while other molecules do not have that function, because they all bounce against the upper side of the container sometimes. Hence, the gas molecules do not have a physical role.

## 3.7  Evaluating Explanatory Practices

### 3.7.1  Introduction

In this section we add a normative component to our toolbox by proposing *clusters of evaluative questions*. The questions are evaluative because each of them somehow asks whether the scientists are "doing the right thing". Each cluster is characterised by a theme, and thus contains evaluative questions that are different but related. The theme of the first cluster (Sect. 3.7.2) is "explanation-seeking questions": the evaluative questions are about the explanation-seeking questions of the scientist(s) whose practice we are analysing. The themes of the other clusters are "format" (Sect. 3.7.3), "ontological level" (Sect. 3.7.4), "level of abstraction" (Sect. 3.7.5) and "irrelevant criteria" (Sect. 3.7.6).

## 3.7.2 Explanation-Seeking Questions and Epistemic Interests

This section starts from a quite trivial observation: not all why-questions we can come up with are equally interesting. This observation motivates our first cluster of evaluative questions:

(EQ1)  What are the interesting explanation-seeking questions about this phenomenon?
What are the interesting explanation-seeking questions in this domain?
Does this scientist ask an interesting explanation-seeking question?
Do scientists in this discipline ask interesting explanation-seeking questions?

Two of these questions are specific (one phenomenon, one scientist), while the two other are more general. An exhaustive account (covering all scientific disciplines, types of phenomena and types of questions) of what counts as an "interesting" explanation-seeking question cannot be given here.[8] However, it is possible to give some positive guidelines. These guidelines have the following form:

If conditions $c_1, \ldots, c_n$ hold, then question X is an interesting explanation-seeking question.

This means that the guidelines specify sufficient conditions for interestingness. We certainly do not claim that the rules we present are jointly exhaustive. One of the reasons why we don't claim exhaustiveness is that we focus on contrastive questions about particular facts. Similar guidelines can be formulated for other types of questions.

Here is the first guideline:

Suppose that object $x$ has property P at time $t$. Then the question "Why does $x$ have property P, rather than the ideal property P*?" is an interesting explanation-seeking question.

P and P* are mutually exclusive properties. We call questions of the type considered in this guideline I-type questions because they invoke an ideal state. Before we give an example and justify this guideline, let us present a second one which is closely related:

Suppose that object $x$ has property P at time $t$ and object $y$ (which like x belongs to class C) has the ideal property P* at $t$. Then the question "Why does $x$ have property P, while $y$ has the ideal property P*?" is an interesting explanation-seeking question.

---

[8] Such an exhaustive account, if possible at all, would require a full length monograph in itself. However, in Weber, Gervais & Van Bouwel (ms.) the issue of interesting versus non-interesting explanation-seeking questions is discussed in more detail. Among other things, this paper contains more examples of guidelines.

We call this I′-type questions because (like the I-type) they invoke an ideal state but in a different way.

We now give an example of each type. Suppose there are two fields of potatoes (*a* and *b*) which are regularly infected with late blight (*Phytophtora infestans*, the oomcyete or microorganism that caused the 1845 potato famine in Ireland). About one of these fields, a researcher might pose an I′-type question:

Why has field *a* been struck by late blight, rather than remained healthy?

This type of question contrasts an observed state of affairs with an ideal one, i.e. one we consider to be preferable. According to our first guideline, it is an interesting explanation-seeking question. Suppose now that suddenly, during harvest time, it is found that the crops of field *b* have remained healthy, while field *a* is struck by late blight. In this case, researchers might pose an I′-type question:

Why has field *a* been struck by late blight, while field *b* remained healthy?

This question contrasts two observed states of affairs, one of which is the ideal one. The two objects have something in common: they both are crop fields. According to our second guideline, this is an interesting explanation-seeking question.

Let us now turn to the justification of the guidelines. The two guidelines we have presented here follow from the fact that *improvement* is an *epistemic interest*. Epistemic interests are types of motives for (a) scientists to search for explanations and (b) other people (policy makers, the general public) to be interested in the explanations scientists give and to pay them for their research. Improvement in the sense of making a given situation better is certainly one of these typical motives. *Prevention* (which is closely related to improvement, but more future-oriented) is another one. A third one is *attributing moral and/or legal responsibility*: this is often the underlying epistemic interest when we explain actions (see Weber and Vanderbeeken 2005 for this).

### 3.7.3 The Appropriate Format of an Explanation

The fact that explanations can have different formats leads to the second cluster of evaluative questions:

(EQ2) Does the answer which this scientist gives to his/her explanation-seeking question have an appropriate format?
Do scientists in this discipline give answers that have an appropriate format?
If scientists choose a format of explanation that does not seem optimal, do they have a good reason for doing this?

The importance of the first two questions is straightforward, given the variety of explanation formats we have presented in Sect. 3.3–3.6 and the links these formats have with types of questions (the format has to be adequate for the type of question at hand). Examples are given in Chap. 4 (e.g. in Sect. 4.2.4). The third question results from the possibility of suboptimal knowledge: it can happen that an explanation of the best possible type cannot be constructed because our current knowledge is not sufficient for doing this. We will discuss such a case in Sect. 4.5.

### 3.7.4 Explanations and Levels of Reality

Our third cluster originates from the fact that explanations can by situated at different ontological levels. Consider the following question:

> Why do Belgians who spend their summer holidays in the Mediterranean area develop skin cancer more often than those who spend their holidays in Belgium?

A possible answer is that in the group of the Mediterranean holiday spenders the exposure to sun rays is higher. We call this answer an *environmental explanation*, because it refers to a property which the whole group has in common and that is external to the individuals in the group. In particular, it refers to the environment of the group as a whole, which stands apart from the properties of the particular individuals within that group. Consider now another question:

> Why do some people with high exposure to sun rays develop cancer of the skin, while others do not?

A possible answer is that some people protect themselves by using a sufficient amount of sun tan lotion, while others don't. We call this answer an *individualist explanation* because it refers to differences among individuals in the group—these can be differences qua behaviour, biological constitution, etc.

This example illustrates that explanations can be situated at different levels. It is important that scientists are aware of this, and take well-informed and well-thought decisions about the level of reality at which they will operate. This leads to the following cluster of evaluative questions:

(EQ3)   Is this scientist considering explanations at different levels of reality and does he/she make a justified decision about the level to be used?
Do scientists in this discipline consider explanations at different levels of reality and do they make justified decisions about the level to be used?

In Sect. 4.6 we give an elaborate example (the Cuban Missile Crisis) in which this cluster is illustrated. Other examples can be found in Van Bouwel et al. (2011).

### 3.7.5 Abstraction and Amount of Detail in Explanations

#### 3.7.5.1 The Kairetic Account of Michael Strevens

The cluster in this section is inspired by Michael Strevens' book *Depth. An Account of Scientific Explanation* (2008). The *kairetic account* which he elaborates in this book is not a new theory about the format of explanations: Strevens uses a causal variant of the DN-model, similar to what Hausman proposes (see Sect. 1.4.1) and to our CDN format in Sect. 3.3.2. What makes Strevens' book important is that it contains an elaborate account of *explanatory relevance*. Let us clarify this. Strevens considers and rejects *the minimal causal account of event explanation*. According to this account …

> [A]n event is explained by whatever other events causally influence it, together with the laws and background conditions in virtue of which they do so (2008, p. 41).

His argument against this view is:

> The most obvious difficulty facing the minimal causal account is the apparently unreasonable vastness of a complete causal explanation. As I pointed out above, in a quasi-Newtonian world like our own, an event's minimal explanation ought in principle to mention anything that has ever exerted a gravitational force on the objects involved in the event, anything that had previously exerted a force on these exerters, and so on. But all scientific explanations, even the most well regarded, describe much less than the complete causal history of the explanandum (2008, p. 43).

Thus the question arises: how do we decide which causal information to include in the explanation and which information to exclude? In order to answer this question, Strevens develops an optimizing procedure which is the core of his kairetic account. We cannot explain this procedure in detail here (see Chap. 4 of his book for this). However, it is important to have an idea of what this procedure does:

> The kairetic theory provides a method for determining the aspects of a causal process that made a difference to the occurrence of a particular event. The essence of the theory is a procedure that does the following: given as input a causal model $M$ for the production of an event $e$, the procedure yields as output another causal model for $e$ that contains only elements in $M$ that made a difference to the production of $e$. A model that contains explanatory irrelevancies is then "distilled" so that it contains only explanatorily relevant factors. (2008, p. 69)

The need for such a procedure has been recognised by many philosophers before Strevens, e.g. Peter Lipton:

> Suppose that my car is belching thick, black smoke. Wishing to correct the situation, I naturally ask why it is happening. Now imagine that God (or perhaps an evil genius) presents me with a full Deductive-Nomological explanation of the smoke. This may not be much help. The problem is that many of the causes of the smoke are also causes of the car's normal operation. Were I to eliminate one of these, I might only succeed in making the engine inoperable. By contrast, an explanation of why the car is smoking rather than running normally is far more likely to meet my diagnostic needs. (1993, p. 53)

So Lipton agrees that a good explanation is one in which explanatorily irrelevant details are weeded out. Strevens provides a procedure for doing this in a systematic way.

An obvious move in optimisation is that we leave out factors that did not make a difference. Strevens calls this *elimination* of causal factors. However, Strevens also discusses the possibility of *abstraction*:

> I throw a cannonball at a window, and the window breaks. Does the fact that the cannonball weighs exactly 10 kg make a difference to the window's breaking? The natural answer to this question is no. The fact that the cannonball is rather heavy made a difference, but the fact that it weighed in at exactly 10 kg did not. (2008, p. 96)

In such cases, the optimizing procedure ensures that the proposition "'The ball's mass was 10 kg" is not in the explanation, but that e.g. the proposition "The ball's mass was greater than 1 kg" is in it. This is what Strevens calls abstraction (2008, p. 97). Abstraction is more refined than elimination because it is gradual, while elimination is all or nothing.

### 3.7.5.2  Another Reason to Leave Out Details

The desideratum of keeping only causal factors that make a difference is certainly plausible in cases where we want to answer a contrastive question, such as in Lipton's example. This is also what Strevens has in mind in his 2008 book, though he does not formulate the questions in a contrastive way. However, it is important that we also look at this issue from a different perspective, viz. resemblance questions. The social revolutions example of Sect. 3.5.3 can illustrate this. The complete causal story about any of the three revolutions contains claims that are irrelevant for explaining the other revolutions because they point at differences rather than at common causal factors. For instance, what is usually called *the* Russian Revolution (October 1917) was preceded by a revolution in February which ended the regime of the tsars but was not a big social revolution. This information is irrelevant for answering the resemblance question. In order to explain the resemblance, we have to focus on the factors that were present in all three cases: external military/economic pressure and a strong sense of community.

### 3.7.5.3  A Cluster of Evaluative Questions

It is important that scientists are aware of the issue of elimination and abstraction and take well-informed and well-thought decisions about the amount of detail they include in their explanations. This leads to the following cluster of evaluative questions:

(EQ4)   Is the level of detail of the explanation adequate?
        Does the theory on which the explanation is based use the right kinds?

The first question is straightforward given what we have said above. The second question is a derivative one. If the theory on which the explanation is based makes too fine-grained distinctions when assorting objects into kinds, the explanation which is based on it will contain irrelevant details. Section 4.7 contains an elaborate example (Robert Merton's analysis of propaganda instruments during World War II) in which this cluster is illustrated. Here we give a brief example to clarify the issue about kinds. Little (1995) discusses an economic model developed by Lustig and Taylor (1990) which is meant to predict the effects (on income, trade balance and government deficit) of certain possible policy interventions (income transfers, price subsidies) in Mexico. In order to do this, they divide the Mexican economy into eight sectors: corn and beans, other agriculture, petroleum, fertilisers, food processing, other industry, services and commerce. This is more fine-grained than the classical fourfold distinction (agriculture—industry—services—commerce). They do this because they think that their fine-grained distinctions make sense: they think there are relevant differences between these sectors. Analogously, they distinguish seven income groups because they think there are relevant differences in the way their income is generated: (independent) peasants, agricultural workers, agricultural capitalists, urban workers, urban capitalists, merchants and urban marginals.

## 3.7.6  Irrelevant Preferences and Mistaken Exclusions

A potential threat to the adequacy of explanations produced by scientists is that they use *irrelevant preference rules*: they think that a property Z is a reason to prefer an explanation X over an explanation Y if X has Z and Y not, while it can be argued that having Z or not is irrelevant to the quality of the explanation. An elaborate example of this will be given in Sect. 4.8.

A similar threat is that scientists may use *irrelevant exclusion rules*: they think that a property Z is a reason to reject an explanation X if X has Z, while it can be argued that having Z is not problematic at all. For instance, in debates on explanation in the social sciences, some philosophers and social scientists have argued that functional explanations or structural explanations are not adequate. We use functional explanations as an example. In a functional explanation we explain a social structure or phenomenon by giving its function or purpose, which is an explanation in terms of its effects rather than by its causes. For example, it might be claimed that the explanation for a certain social practice in a society is the way in which it contributes to social stability or group solidarity. Some philosophers of social science have claimed that functional explanation is not a good practice. For instance, Elster writes that "there is no place for functional explanation in the social sciences" (1984, p. viii). Other philosophers of social science have claimed that this exclusion is not well-founded (see Van Bouwel and Weber 2008a, b).

The possibility of irrelevant preference rules and irrelevant exclusion rules leads to the following cluster of evaluative questions:

(EQ5)    Does this scientist use irrelevant preference rules?
         Do scientists in this discipline often use irrelevant preference rules?
         Does this scientist mistakenly exclude certain types of explanation?
         Do scientists in this discipline often mistakenly exclude certain types of explanation?

In Sect. 4.8 we present an elaborate example which illustrates this cluster.

## References

Bechtel W, Abrahamsen A (2005) Explanation: a mechanist alternative. In: Studies in history and philosophy of biological and biomedical sciences, vol 36, pp. 421–441
Craver C (2007) Explaining the brain. Clarendon Press, Oxford
Elster J (1984) Ulysses and the Sirens. Cambridge University Press, Cambridge

Elster J (2007) Explaining social behavior. More nuts and bolts for the social sciences. Cambridge University Press, Cambridge

Friedman M (1974) Explanation and scientific understanding. J Philos 71:5–19

Giere R (1997) Understanding scientific reasoning, 4th edn. Harcourt Brace College Publishers, Fort Worth

Hausman DM (1998) Causal asymmetries. Cambridge University Press, Cambridge

Halonen I, Hintikka J (1999) Unification—it's magnificent but is it explanation? Synthese 120:27–47

Hempel C (1965a) Aspects of scientific explanation and other essays in the philosophy of science. Free Press, New York

Hempel C (1965b) Aspects of scientific explanation. In: Hempel C (1965a), pp 331–496

Hempel C, Oppenheim P (1948) Studies in the logic of explanation. In: Hempel C (ed.) (1965a), Philos Sci 15:245–290

Holton G, Roller D (1958) Foundations of modern physical science. Addison-Wesley, Reading

Kitcher P (1981) Explanatory unification. Philos Sci 48:507–531

Kuipers T (2001) Structures in science: heuristic patterns based on cognitive structures. Kluwer, Dordrecht

Lipton P (1993) Making a difference. Philosophica 51:39–54

Little D (1995) Economic models in development economics. In: Little D (ed) The reliability of economic models. Kluwer, Boston, pp 243–270

Lustig N, Taylor L (1990) Mexican food consumption policies in a structuralist CGE model. In: Taylor L (ed) Socially relevant policy analysis. MIT Press, Cambridge

Machamer P, Darden L, Craver C (2000) Thinking about mechanisms. Philos Sci 67:1–25

Morgan S, Winship C (2007) Counterfactuals and causal inference. Methods and principles for social research. Cambridge University Press, Cambridge

Skipper RA (1999) Selection and the extent of explanatory unification. Philos Sci 66:S196–S209

Skocpol T (1979) States and social revolutions. Cambridge University Press, Cambridge

Strevens M (2008) Depth: an account of scientific explanation. Harvard University Press, Cambridge

Taylor M (1988) Rationality and revolutionary collective action. In: Taylor M (ed) Rationality and revolution. Cambridge University Press, Cambridge, pp 63–97

Van Bouwel J, Weber E (2008a) De-ontologizing the debate on social explanations: a pragmatic approach based on epistemic interests. Hum Stud 31:423–442

Van Bouwel J, Weber E (2008b) A pragmatic defence of non-relativistic explanatory pluralism in history and social science. Hist Theory 47:168–182

Van Bouwel J, Weber E, De Vreese L (2011) Indispensability arguments in favour of reductive explanations. J Gen Philos Sci 42:33–46

Van Fraassen B (1980) The scientific image. Oxford University Press, Oxford

Weber E, Geravis R, Van Bouwel J (ms.) The 'green cheese' and 'red herring' problems reconsidered. Epistemological vs. methodological tasks for philosophers of science

Weber E, Van Bouwel J (2009) Causation, unification and the adequacy of explanations of facts. Theoria 66:301–320

Weber E, Vanderbeeken R (2005) The functions of intentional explanations of actions. Behav Philos 33:1–16

Weber E, Van Bouwel J, Lefevere M (2012) The role of unification in explanations of facts. In: De Regt H, Okasha S, Hartmann S (eds) EPSA philosophy of science: Amsterdam 2009. Springer, Dordrecht, pp 403–413

Weber E, Van Bouwel J, Vanderbeeken R (2005) Forms of causal explanation. Found Sci 10:437–454

Weber E, Van Dyck M (2002) Unification and explanation. A comment on Halonen, Hintikka and Schurz. Synthese 131:145–154

Woodward J (2003) Making things happen. A theory of causal explanation. Oxford University Press, New York

Wouters A (2003) Four notions of biological function. Stud Hist Philos Biol Biomed Sci 34:633–668

# Chapter 4
# Examples of Descriptions and Evaluations of Explanatory Practices

## 4.1 Introduction

The examples we gave in Chap. 3 are brief and not always realistic. In this chapter we present elaborate and realistic examples of how the toolbox of Chap. 3 can be used for analysing explanatory practices.

Sections 4.2 and 4.3 contain examples of *descriptions* of explanatory practices. In Sect. 4.2 we give an example of how the toolbox can be used to describe the explanatory practice of an *individual scientist*: we analyse the explanation-seeking questions and answers of Richard Feynman during his investigation of the explosion of the Challenger space shuttle. In Sect. 4.3 we give an example of how the toolbox can be used to describe the practice of a *discipline* (or an important group of scientists within a discipline): we will look at explanations in dynamical cognitive science.

Sections 4.4–4.8 contain examples of *evaluations* of explanatory practices. Each of these sections is devoted to one of the clusters introduced in Sect. 3.7 of Chap. 3. In Sect. 4.4, we look at explanations of actions in order to illustrate cluster (EQ1). Section 4.5 is about pigeon navigation and illustrates cluster (EQ2). In Sect. 4.6 we discuss explanations related to the Cuban Missile Crisis in order to illustrate cluster (EQ3). Section 4.7 uses Robert Merton's analysis of propaganda during World War II to illustrate cluster (EQ4). Finally, in Sect. 4.8 we illustrate cluster (EQ5) by means of the debate on proximate and remote causes in explanations in the social sciences.

## 4.2  Richard Feynman on the Challenger Disaster

### 4.2.1  Introduction

The Challenger disaster happened at Kennedy Space Center in Florida on January 28, 1986.[1] Space shuttle Challenger lifted off at 11:38:00 a.m. (Eastern Standard Time) and exploded 73 s later. The entire crew (seven astronauts) was killed. To investigate the causes of the accident, president Reagan appointed a presidential commission. It was headed by former secretary of state William Rogers and included, among others, former astronaut Neil Armstrong and Nobel Prize winning physicist Richard Feynman.

Feynman tells about this episode in his life in one of his books (Feynman 1988). Feynman and the commission as a whole focussed on two questions. The first was: what physically caused the explosion of the Challenger? The second question was: what went wrong in NASA that made the explosion possible? Before we have a closer look at Feynman's search for information and the resulting explanations, it is useful to give some background knowledge. The fatal mission (its official name was mission 51-L) was in several respects a standard mission. Challenger was not the first operational space shuttle (Columbia was the first one) and it had done nine successful missions, so there was nothing exceptional in those respects. Also, the cargo was the usual stuff (a satellite and scientific apparatus). However, the mission was exceptional in two respects. First, one of the crew members was Christa MacAuliffe, who was supposed to become the first school teacher in space. This caused large media coverage of the accident. Second, it was a very cold day. This turned out to be important, as we will see below.

### 4.2.2  An Etiological Explanation

To find an answer to the first question, Feynman obtained technical information, primarily on the working of the solid rocket boosters (SRBs), the space shuttle main engines (SSMEs) and the flight electronics (see Feynman (1988) for details). Very early in his search for answers, it became clear to Feynman that there was a serious problem with the rubber O-rings that should seal the aft field joint of the right SRB. SRBs are built up from different smaller segments that are connected by a Tang and Clevis joint: each segment has a U-like shape at the top, such that the bottom of the segment above can nicely slide into it. To prevent that the gases inside the SRB can leak through these joints, they are sealed with two rubber O-rings that are supposed to be very flexible. However, it turned out that it was not clear how these rubber seals would react to the extremely cold temperature on the

---

[1] There are various online resources which describe what happened on that day. We use the texts of Greene 2012a and 2012b.

morning of the launch, and that some technicians had already expressed their fear for problems resulting from the O-rings beforehand.

What Feynman actually did to get an answer to his first question, was searching for an acceptable etiological explanation (see Sect. 3.3.5) which described the detailed technical process that led to the explosion. He investigated what technicians knew about the behaviour of the specific parts of the space shuttle's rocket boosters and main engines, and about the influence of the specific conditions under which these mechanisms had to work on the specific morning of the Challenger's launch, namely the extreme cold. These brought him to an explanation showing how gases could leak, causing a fire blowing up the space shuttle. Nick Greene summarises this explanation as follows:

> The commission's report cited the cause of the disaster as the failure of an "O-ring" seal in the solid-fuel rocket on the Space Shuttle Challenger's right side. The faulty design of the seal coupled with the unusually cold weather, let hot gases to leak through the joint. Booster rocket flames were able to pass through the failed seal enlarging the small hole. These flames then burned through the Space Shuttle Challenger's external fuel tank and through one of the supports that attached the booster to the side of the tank. That booster broke loose and collided with the tank, piercing the tank's side. Liquid hydrogen and liquid oxygen fuels from the tank and booster mixed and ignited, causing the Space Shuttle Challenger to tear apart. (Greene (2012b))

## 4.2.3  A Contrastive Etiological Explanation

This causal–mechanical explanation, although giving us good insight in what caused the explosion of the Challenger, does not answer the second explanation-seeking question that Feynman and the committee addressed. However, it enabled them to formulate two follow-up questions:

> Why was the space shuttle program continued, rather than put on hold till the problems of gas leakage and erosion were solved?
>
> Why was the decision taken to launch the Challenger on January 28th, instead of postponing the launch to a less cold day?

These questions are more specific than the original second question (What went wrong within NASA?) They presuppose knowledge of what physically caused the explosion. The answers which Feynman and the presidential committee give are contrastive etiological (see Sect. 3.4.2): they first describe how NASA really functioned and then tell us how things should have been different in order to obtain the alternative fact. We focus on the first follow-up question.

During some previous missions there had been gas leakages in some seals, resulting in clearly visible black marks on the rocket (at places where hot gas had leaked) and so-called 'erosion' (spots where the O rings were partially burnt). These two problems show that the O-rings did not perform their function (viz. sealing) perfectly. This shortcoming is the "faulty design" to which Greene refers in the quote above. According to Feynman, the way in which NASA officials dealt with these incidents was completely wrong. He compares their risk analysis methods and risk management strategies to Russian roulette. Here is what Feynman writes:

> The phenomenon of accepting for flight, seals that had shown erosion and blow-by in previous flights, is very clear. The Challenger flight is an excellent example. There are several references to flights that had gone before. The acceptance and success of these flights is taken as evidence of safety. *But erosion and blow-by are not what the design expected. They are warnings that something is wrong.* The equipment is not operating as expected, and therefore there is a danger that it can operate with even wider deviations in this unexpected and not thoroughly understood way. The fact that this danger did not lead to a catastrophe before is no guarantee that it will not the next time, unless it is completely understood. When playing Russian roulette the fact that the first shot got off safely is little comfort for the next. The origin and consequences of the erosion and blow-by were not understood. They did not occur equally on all flights and all joints; sometimes more, and sometimes less. Why not sometime, when whatever conditions determined it were right, still more leading to catastrophe?
>
>   In spite of these variations from case to case, officials behaved as if they understood it, giving apparently logical arguments to each other often depending on the "success" of previous flights. For example. in determining if flight 51-L was safe to fly in the face of ring erosion in flight 51-C, it was noted that the erosion depth was only one-third of the radius. It had been noted in an experiment cutting the ring that cutting it as deep as one radius was necessary before the ring failed. Instead of being very concerned that variations of poorly understood conditions might reasonably create a deeper erosion this time, it was asserted, there was "a safety factor of three." This is a strange use of the engineer's term,"safety factor." If a bridge is built to withstand a certain load without the beams permanently deforming, cracking, or breaking, it may be designed for the materials used to actually stand up under three times the load. This "safety factor" is to allow for uncertain excesses of load, or unknown extra loads, or weaknesses in the material that might have unexpected flaws, etc. If now the expected load comes on to the new bridge and a crack appears in a beam, this is a failure of the design. There was no safety factor at all; even though the bridge did not actually collapse because the crack went only one-third of the way through the beam. The O-rings of the Solid Rocket Boosters were not designed to erode. *Erosion was a clue that something was wrong. Erosion was not something from which safety can be inferred.*
>
>   There was no way, without full understanding, that one could have confidence that conditions the next time might not produce erosion three times more severe than the time before. Nevertheless, officials fooled themselves into thinking they had such understanding and confidence, in spite of the peculiar variations from case to case. (Feynman 1986; italics added)

This quote nicely summarizes the descriptive part of Feynman's explanation (there are much more details about specific decisions and measurements in other parts of the Commission report). The sentences we put in italics point at the different assessment procedure, which in Feynman's view, would have lead to the alternative result: NASA officials should have interpreted the gas leakages and erosions as warning signs, as indications that something was wrong.

## 4.2.4  Summary and Some Brief Evaluative Remarks

Our findings can be summarised as follows. The first question that was put forward at the beginning of the investigation was a plain explanation-seeking question about a particular fact. It was answered by means of an etiological explanation. The second initial question was replaced with more specific ones as soon as the first question was answered. The answers to the follow-up questions were contrastive etiological.

The main aim of this section is to use the toolbox in order to describe the explanatory practice of Richard Feynman and his colleagues in the committee. However, we add some brief evaluative remarks which relate to clusters (EQ1) and (EQ2) introduced in Sect. 3.7. The questions in Sect. 4.2.3 were surely interesting: answering them is useful for preventing similar accidents. The initial questions are interesting because they lead to these follow-up questions. So the scientists were asking the right questions (cf. (EQ1)). The way they answer the questions is appropriate (cf. (EQ2)): an etiological explanation for the plain question, contrastive etiological explanations for the contrastive questions.

## 4.3  Explanations in Dynamical Cognitive Science[2]

### 4.3.1  Introduction

In Walmsley (2008), it has been argued that the explanations employed in the dynamical approach to cognitive science, as exemplified by the HKB model of rhythmic finger movement (Haken et al. 1985) and the model of infant perseverative reaching developed by Esther Thelen and her colleagues (Thelen et al. 2001), conform to the DN model. Walmsley's approach is the one we advocated in Chap. 2. More specifically, he uses philosophical models of scientific explanation in order to describe the explanatory practices of a group of cognitive scientists. So we think he is on the right track metaphilosophically: his approach is sound. However, we think there are two problems with his results. First, by characterizing explanations in dynamical cognitive science as deductive-nomological, Walmsley neglects an important property of the explanations, viz. that they are causal. In this way he suggests that they are problematic (cfr. the problems with Hempel's models), while they are not. Second, not all explanations in dynamical cognitive science are deductive-nomological. At least some of them fit the CDR model, i.e. the non-deductive variant of the covering law model described in Sect. 3.3.6 which uses default rules instead of strict, exceptionless laws. The structure of this section is straightforward: we briefly sketch Walmsley's position, and then discuss these two problems.

### 4.3.2  Walmsley's Claim

Why does Walmsley think that "dynamical explanations are covering law explanations" (2008, p. 346)? He draws on two separate strands of evidence to support this claim.

---

[2] This section is based on Gervais and Weber (2011).

The first strand of evidence comes from scientific practice; two models used in dynamical cognitive science are discussed. The first model, the so-called HKB model of rhythmic finger tapping, attempts to explain the curious observation that test subjects, having placed their hands palm-down on a table, can oscillate both index fingers in 'phase motion' (to the left and right at the same time) reliably across higher frequencies than they can oscillate them in 'antiphase motion' (to the left with one finger while to the right with the other). Dynamical systems theory explains this fact by capturing the rate of change of relative phase, the periodic function of current relative phase and the frequency of oscillation in a mathematical equation. Thus, the dynamicist is able not only to describe behaviour that has already been observed, but also to predict behaviour that can be (and indeed, has been) confirmed by subsequent experiments. The second model Walmsley considers is Thelen et al.'s model of infant perseverative reaching. This model attempts to explain the 'A-not-B error': a child between 7 and 12 months old is presented with two boxes. When an adult comes in and hides a toy or piece of candy under one of the boxes, the child will reach for the correct box. But if the adult repeats this procedure several times and then suddenly hides the toy under the other box, the child will still reach for the first box, even though it has observed the adult hiding the toy under the other box. Dynamical systems theory explains this fact by means of an equation, relating the current state of the movement planning field, general and specific aspects of the task, etc. Walmsley notes that a particular pattern of behavior in this kind of experimental setup follows as a mathematical and deductive consequence of the equation in conjunction with the initial states, and has the same logical form as the prediction of that event would have taken.

The features of these explanations (explanandum as a logical consequence of the explanans, equivalence of explanation and prediction) put them squarely in the covering law model. From this Walmsley draws the conclusion that "… some dynamical models provide covering law explanations" (2008, p. 342).

The second strand of evidence is meant to support the conclusion that dynamical cognitive scientific explanation will keep on producing covering law explanations (so the current state will be preserved). This evidence is drawn from a consideration of the *goals* which dynamical cognitive scientists set themselves. Here, Walmsley quotes a number of eminent theorists and philosophers who concur that the aim of dynamical modeling is to provide explanations that deduce the explananda from equations and certain given parameters, and contends that "… these quotations, coupled with the lack of an alternative metatheoretical stance toward dynamical explanation, are sufficient to establish that the explanatory goal of dynamical cognitive scientists is to provide covering law explanations…" (2008, p. 343).

### 4.3.3  Causal Explanations

The explanations presented by the authors which Walmsley discusses are causal. Thelen et al.'s model of infant perseverative reaching explains an infant's reaching behaviour as a mathematical consequence of an equation plus a number of values for the parameters and variables, including the current state of the movement field, the general specific and memory inputs to the system and a function integrating competing inputs. From the perspective of the traditional covering law model, there is nothing against reversing the order of the argument: from the equation, parameters and variables, together with the reaching behavior of the child, we might deduce (hence 'explain') under which box the researcher has hidden the toy. As we have seen in Sect. 1.3.3, many critics of Hempel claim that this kind of argument has to be ruled out as explanation. And what is important for us: Thelen et al. do not claim that such arguments are explanations. They only claim that the argument which starts from the causes and has the effect as conclusion is an explanation. Similarly, the authors of the HKB model do not claim that effects can explain their causes.

The upshot of this is that we can give a more precise characterisation of the examples: they fit the CDN format we have developed in Sect. 3.3.2. This is important because, if they would not fit this model, dynamical cognitive science could have been accused of providing pseudo-explanations like the one in which the height of a flagpole is explained by the length of its shadow, instead of genuine explanations. Walmsley acknowledges the validity of the counterexamples raised against the traditional covering law model. In his view they "… show […] that explanations in dynamical cognitive science will be  subject to the same set of criticisms, in virtue of the form they take" (2008, p. 344). This conclusion is not correct: the explanations have a more specific form than Walmsley admits, and therefore do *not* share the problems of Hempel's DN model.

## 4.3.4  Non-Deductive Explanations

In Hempel and Oppenheim's (1948) paper, "covering-law explanation" and "deductive-nomological explanation" are used as synonyms. Later on Hempel distinguished between deductive-nomological explanations and inductive-statistical explanations and used "covering-law explanation" as an overarching label. Walmsley does not consider IS explanations. He only discusses Hempel's DN model (2008, p. 338–340) and claims that the explanations of dynamical cognitive science fit this model. He explicitly says that in the explanations of Thelen et al. and in the HKB explanations the explanandum is a *deductive* consequence of the explanans (pp. 340–341). This is strange, because his paper also contains information from which it follows that this claim is not correct. When describing the effect explained by Thelen et al., he mentions that it is "enormously sensitive to slight changes in the experimental conditions, such as the delay between viewing and reaching, the way the scene is viewed, the number of trials, the presence of distracting stimuli, and so on" (p. 335). The model of Thelen et al. is superior to previous models because it can take into account many of these contextual subtleties. However, as long as the model cannot cope with *all* contextual variation, it cannot produce deductive-nomological explanations. The reason is that, if the model cannot account for all contextual variation, the law we can derive from it has the form "If initial conditions $C_1$, $C_2$, ... $C_N$ are satisfied, then *usually* E happens". In order to have deductive-nomological explanation, we need "always" in the law instead of "usually": DN explanations need strict, exceptionless laws of the form "If initial conditions $C_1$, $C_2$, ... $C_N$ are satisfied, then E *always* happens".

The law we actually can derive from Thelen et al.'s model has the format of a default rule, as clarified in Sect. 3.3.6. Their explanations fit our CDR model. They are causal, as established in the previous section. They use a covering law and have the form of an argument (this has been shown by Walmsley). But the covering laws they use have exceptions, so the explanations are not deductive. This means that they either use a default rule (as we suggest) or a probability statement (and thus are causal IS explanations). The latter presuppose that we can give a precise relative frequency of "normal cases" and "exceptions". This does not seem the case in the explanations of Thelen et al., so they should be seen as CDR explanations. As Walmsley himself acknowledges (p. 344), exceptionless generalisations have been hard to find in psychology. So there are reasons to suppose that explanations using default rules are the rule, rather than the exception.

### 4.3.5  Conclusion

In this section, we have examined Walmsley's claim that the explanations in dynamical cognitive science conform to Hempel's DN model. We have shown that the claim is misleading because the explanations of dynamical cognitive science are causal, so they are not as problematic as Walmsley suggests. Furthermore, we have shown that not all explanations in dynamical cognitive science are deductive-nomological. At least some of them (and presumably most of them) fit our CDR model: they use default rules instead of strict, exceptionless laws.

## 4.4  Intentional Explanations of Actions

### 4.4.1  Introduction

In the previous sections we have used the toolbox of Chap. 3 to describe explanatory practices. In Sect. 4.4–4.8 we will focus on the evaluation of explanatory practices. The aim of this section is to give an elaborate example of cluster (EQ1) and answers to the questions in this cluster.

*Actions* as we conceive them transform an indeterminate situation into a determinate one. This terminology is taken from John Dewey (1938). An indeterminate situation is a situation in which we know that *something* must be done, but do not know *what* to do. A determinate situation is a situation in which we know what to do. By limiting ourselves to actions defined in this way, we exclude emotional and habitual behavior from the scope of this section. We do this to keep the discussion clear and short. However, since no restrictions are posed on the decision procedures by which the indeterminate situation is transformed into a determinate one (e.g., flipping a coin is allowed, as well as maximizing expected utility), our analysis is *not* restricted to actions that result from procedures in which the possible consequences of each action are considered. When the agent flips a coin, his choice is made without considering the outcomes of possible actions. When the agent maximizes expected utility, the possible outcomes of actions do influence decisions. We want to include both cases in which an outcome-neglecting rule (such as flipping a coin) is used and cases in which an outcome-oriented rule (such as expected utility maximizing) is used. For actions conceived in this way, we can ask I-type questions (and often do so). In Sects. 4.4.2 and 4.4.3 we investigate why such questions are useful.

## 4.4.2 Explanation-Seeking Questions About Actions[3]

Consider a scientist who has decided that he and the research group he is directing, will co-operate in the development of new weapons. Then we can ask the following question[4]:

> Why did this scientist decide to co-operate in the development of new weapons, rather than refrain from this?

This is an I-type question about an action. An answer to this question can be interesting for several reasons:

(a) The explanation may help us to determine our moral judgment about the decision (was the scientist's decision morally justified?).
(b) Assuming that withdrawal is still possible and we disapprove the decision, the explanation might teach us how to make the scientist withdraw.
(c) Assuming that we disapprove, the explanation might teach us how we can avoid that the same or other scientists volunteer for similar research programs in the future.

These reasons illustrate the functions that answers to contrastive questions about actions may have: a *judgment determining function* (they may help us to make moral or legal judgments), a *therapeutic* function (they may diagnose what went wrong and thus help us to restore an ideal state) and a *preventive* function (they may help us to avoid similar actions by other people on future occasions).

---

[3] This section is based on Weber and Vanderbeeken (2005).

[4] This question is inspired by Lackey (1994); see Sect. 4.4.3 for details.

### 4.4.3 A Case Study

We will now discuss these functions in detail. The material for the example we use is taken from Lackey (1994). The aim of Lackey's paper is ethical: he wants to argue against participation in military research programs. But his argument is based on explanations of (imaginary) actions, so we can use his material here. Lackey considers an imaginary scientist leading a group of researchers who decides to take funds budgeted through the Strategic Defense Initiative (SDI). According to Lackey, three cases can be distinguished:

(1) The scientist thinks his research will not serve any military purpose, because the goals of SDI cannot be achieved.
(2) The scientist believes that his research will serve some military purpose, and that this purpose is evil.
(3) The scientist believes that his research will serve some military purpose, and believes that this purpose is benign.

In all three cases, there is a discrepancy between the actual preference forming process of the scientist and the presumed ideal pattern. The latter may be characterized as follows (we are aware that not everyone will agree that this is the ideal pattern; however, this is the ideal pattern on which Lackey bases his contrastive explanations):

(1') The scientist should know that research towards unattainable goals is often diverted towards other military purposes (a causal belief).
(2') The scientist should be ready to avoid evil (a moral principle).
(3') The scientist should know that the use of force by the US since the Korean war is mostly motivated by economic and geopolitical self-interest (a factual belief).

In case (2), where the discrepancy is that the scientist does not conform to (2'), our judgment is clear: the act of the scientist is immoral. In the other cases, where the scientist has false beliefs so that (1') or (3') is not satisfied, our judgment depends on whether the ignorance is acceptable. These cases are less straightforward: it depends on the minimal efforts in information gathering we require for a moral decision. The discrepancies revealed by the explanations determine our moral judgment, together with other factors, such as what counts as acceptable ignorance and what as culpable ignorance.

The example shows that the (contrastive) explanations we construct as answers to I-type questions about actions can have a *judgment-determining* function. In our example the judgment is of a moral order: the question to be decided is whether the action was morally justified or not. In other cases, the question may be of a legal order (e.g. is this person guilty of murder or not?) or may relate to status (e.g. should president X resign from his job for what he has done?). The answer to such questions also depends on the discrepancies between the real world and the ideal pattern we (i.e. the explainers) have in mind.

With respect to the second function (restoring the ideal state) it may be hard to change someone's moral principles (second discrepancy), but changing someone's factual beliefs may be easier. To the extent that these factual beliefs are typical for scientists working with military funds, explaining the decision of one research group can provide a strategy for avoiding similar actions in comparable groups (cf. third function). According to Lackey historical ignorance (third discrepancy) is typical for groups working with military funds:

> It has not been my impression that many scientists involved in defense work or subsisting on the defense dole have undertaken an examination of the historical evidence and the present world scene. On the contrary, one finds, given the educational attainments of these people, a surprising ignorance of history, and an incredible lack of exposure to alternative interpretations of what happened and why. William Broad's account of the political naiveté of young scientists at the Lawrence Livermore National Laboratory describes not the odd case, but the typical ignorance of history and the humanities one finds among these researchers. (1994, p. 404)

The study referred to is Broad (1985). The example shows that contrastive explanations we construct as answers to I-type questions can have a *therapeutic* function (they diagnose what went wrong and thus help us to restore an ideal state) or a *preventive* function (they help us to avoid similar actions by other people on future occasions).

### 4.4.4 Conclusion

Let us take stock. We have checked whether the question "Why did this scientist decide to co-operate in the development of new weapons, rather than refrain from this?" is an interesting explanation-seeking question. This fits into cluster (EQ1). We have found out that this question may be interesting for three reasons. In the terminology introduced in Sect. 3.7.2, we can say that we have found three *epistemic interests* that may motivate the question: a judgment-determining interest, a therapeutic interest and a preventive interest. Each of these epistemic interests corresponds to one of the possible functions of answers described in Sect. 4.4.3. Indeed, epistemic interests and functions of answers are two sides of the same coin: an epistemic interest is a motivation for asking a question, a function of an answer is a way in which we can use this answer (something we can do with it).

## 4.5 Pigeon Navigation[5]

### 4.5.1 Introduction

This section is about cluster (EQ2) presented in Sect. 3.7.3, with special attention to the possibility of suboptimal knowledge. We will argue that (a) biologists often develop covering law explanations of biological capacities, (b) these covering law

---

[5] This example is taken from Gervais and Weber (ms.).

explanations belong to a specific subtype, and (c) that biologists construct these explanations in cases where, given the current state of scientific knowledge, it is impossible to construct a (potentially more interesting) mechanistic explanation.

### 4.5.2  The Explanations

The examples of covering law explanations in biology we shall consider relate to the capacity of homing pigeons (*Columba livia*) to navigate. The source for this example is Keeton and Gould (1986), pp. 575–585. A trained pigeon can be taken from home, transported over very long distances (hundreds of miles are not uncommon) and still find the way back to its home after being released. Moreover, they are able to exercise this capacity both in sunny weather and on cloudy days. This gives rise to two explananda:

Expl. 1:  Pigeons have the capacity to find their way back home on sunny days
Expl. 2:  Pigeons have the capacity to find their way back home on cloudy days

With respect to the first explanandum, it was shown that pigeon navigation depends on the position of the sun as a reference point. Thus, a covering law explanation of Expl. 1 posits an internal sun compass:

Law 1:  Pigeons have a sun compass
Law 2:  All animals with a sun compass have the capacity to find their way back home on sunny days
_____

Expl. 1:  Pigeons have the capacity to find their way back home on sunny days

As already mentioned above, pigeons that are released on a cloudy day have the capacity to find the way back home, hence Expl. 2. In a set of experiments, William Keeton demonstrated that pigeons also have a magnetic compass. Keeton released birds that had magnets and birds with brass rods (which function as placebos) attached to them, both on sunny and on cloudy days. The results were clear: on sunny days, the birds were unaffected, but on cloudy days the birds carrying magnets became disoriented. In this way, Keeton arrived at the following covering law explanation for Expl. 2:

Law 3:  Pigeons have a magnetic compass
Law 4:  All animals with a magnetic compass have the capacity to find the way back home on cloudy days
_____

Expl. 2:  Pigeons have the capacity to find the way back home on cloudy days

Furthermore, the experiments of Keeton suggest conditions under which the two systems become operative:

Law 5:  Pigeons have a sun compass and a backup magnetic compass that works only on cloudy days

Law 6:   All animals with a sun compass and a magnetic compass that works only
         on cloudy days, have the capacity to find the way back to their house on
         sunny days even if they carry a magnet around their neck
         _____
Expl. 3: Pigeons have the capacity to find the way back to their house on sunny
         days, even if they carry a magnet around their neck

## 4.5.3  Properties of the Explanations

In Sect. 4.5.2 we presented three examples of explanations of biological capacities. Let us now analyze their properties. First, they are covering law explanations,
which means that explanations of this type are at least possible in biology. Second,
they posit the existence of a mechanism without describing it. Let us clarify what
we mean by this. The claim that pigeons have a solar compass is, in our view,
identical in meaning to the following claim:

> In the body of pigeons there are entities (of which we don't know where they are and what
> they look like) that have certain unknown activities and are organized in an unknown way.
> These entities, activities and organization ensure that pigeons have the capacity (on sunny
> days) to determine the angle they have to maintain relative to the sun.

The claim that pigeons have a magnetic compass is in our view identical in
meaning to the following:

> In the body of pigeons there are entities (of which we don't know where they are and what
> they look like) that have certain unknown activities and are organized in an unknown way.
> These entities, activities and organization ensure that pigeons have the capacity (on cloudy
> days) to determine the angle they have to maintain relative to the magnetic field of the earth.

If one agrees that this is the meaning of Law 1 and Law 3, the explanations
are not mechanical (because no information is given about the entities, activities
or organization). However, from an ontological point of view they presuppose a
mechanism: the laws cannot be true unless there is a mechanism. So the covering
law explanations belong to a specific subtype: they posit the existence of a mechanism without describing it. This property is certainly not shared by all covering
law explanations of regularities. The explanations in our subtype are hybrid: they
have the form of a deductive argument (this is the property they take from covering law explanations) and they posit a mechanism (this is the property they take
from mechanistic explanations). Explanations of this subtype do not suffer from
the possible problems for covering law explanations of regularities (e.g. circularity) at which Hempel pointed (see Sect. 3.6.2).

## 4.5.4  Evaluation of the Explanatory Practice

We can ask two (interrelated) evaluative questions here. First, is the covering law format appropriate here? Second, do the scientists have a good reason for not giving a

mechanistic explanation? These questions are instantiations of the questions in (EQ2). The answer to both questions is positive: till today it is impossible to give a mechanistic explanation. So there is an epistemic constraint: mechanistic explanations may be more interesting than the covering law explanations, but they are impossible at this moment. Though the experiments of Keeton date from the 1970s (they were published in *Scientific American* in 1974) it is still impossible to give mechanistic explanations of the capacities. To illustrate this, let us take a closer look at the magnetic compass. A relatively recent proposal is that iron particles (superparamagnetic magnetitie or SPM particles) in the nerve terminals of sensory nerves in the upper beak of the homing pigeons might play a role in the mechanism underlying their capacity to find the way home (Fleissner et al. 2003). The hypothesis is that these particles react to the magnetic field of the earth, so that the nerve cells in the upper beak act as magnetoreceptors, passing on information to the brain, allowing the pigeon to determine its direction, height and location. However, the researchers stress that these SPM particle-clusters in the beak are only a *candidate* for the responsible magnetoreceptor (Fleissner et al. 2003, p. 360). Moreover, their conclusions have been disputed recently by a group of researchers who argue that the cells in which the SPM articles are found, are in fact not nerve cells at all, but rather specialized white blood cells (macrophages), whose function is to recycle iron particles of red blood cells (Treiber et al. 2012). If that is true, then it is implausible that they play a role as magnetoreceptors, as white blood cells do not possess the ability to convey information to the brain. Moreover, the number of SPM cells varies widely among individual pigeon beaks, which seems at odds with their supposed roles as magnetoreceptors. Treiber et al. conclude that "…our work reveals that the sensory cells that are responsible for trigeminally mediated magnetic sensation in birds remain undiscovered. These enigmatic cells may reside in the olfactory epithelium, a sensory structure that has been implicated in magnetoreception in the rainbow trout" (Treiber et al. 2012, p. 369). In short, the message of Treiber et al. is: 'we don't know yet where the magnetic compass is located, and since we have failed with the beak, let's now look at the nasal cavity, because that is where rainbow trout have it'.

## 4.6  The Cuban Missile Crisis[6]

### *4.6.1  Introduction*

The Cuban Missile Crisis is a social phenomenon about which many significant why-questions can be asked. We will focus on two questions:

Why did the USSR put missiles in Cuba?

---

[6]  This section uses material from De Langhe et al. (2007).

Why did the USSR decide to place offensive missiles in Cuba without camouflaging the nuclear sites during construction?[7]

Answers to these and many other questions can be found in Graham Allison's famous books on this topic (Allison 1971, Allison and Zelikow 1999). In Sect. 4.6.2 we introduce the three models for analysing international relations which Allison presents in his book and illustrate what their implementation results in in the case of the Cuban Missile Crisis. In Sect. 4.6.3 we show how these models can be used to answer the two specific explanation-seeking questions above and link this to cluster (EQ3).

## 4.6.2  Allison's Models

The models which Allison presents are the Rational Actor Model (Model I), the Organizational Behaviour Model (Model II) and the Governmental Politics Model (Model III). For each model Allison elaborates a theoretical framework which he then uses to analyse the Cuban Missile Crisis. We start with a summary of the theoretical frameworks and their application to the Crisis.

### 4.6.2.1  Model I: Rational Actor Model

Within this model, international relations are made up of the interplay between unitary nation states that act on a rational basis, i.e. they strive for utility maximization. The selected action, according to this model, is the one with the best cost–benefit ratio. Actions are solutions to strategic problems, and the states are supposed to choose the optimal solution.

Implementation of this model for the Cuban Missile Crisis leads to various claims about the decisions of "the USA" and "the USSR". These two states are treated as indivisible units that make rational decisions. For instance, it is claimed that, in response to the placement of missiles in Cuba the USA picked the option of the blockade rather than an air strike or doing nothing, because that was the rational option, avoiding a quick escalation and leaving the USSR to make the next move.

### 4.6.2.2  Model II: Organizational Behaviour Model

This model opens the black box of the unitary states. In this model, international relations are seen as the result of the interplay between the myriad of organizations constituting the state. Primary inferences in this model follow the logic of

---

[7] The Russians started camouflaging the sites only after U-2 flights had pinpointed their locations.

organization instead of the logic of optimization. State output is no longer aimed at one clear goal, but is the common denominator of a whole set of forces, the result of which might well be something none of the organizations had called for. Furthermore, a typical feature of organizations is that they always strive for bigger budgets. They are also cumbersome. As such they act on the basis of standard operating procedures (SOP) that were devised for earlier purposes instead of reacting on the basis of present challenges. Because of this slow response time, organizations have a strong urge to decrease uncertainty. The alternatives open to an organizational actor are severely limited by its repertoire of SOP's.

Implementation of this model for the Cuban Missile Crisis leads to various claims about relevant organisations within the USA and the USSR. Examples of relevant organisations are for the USA: the US Navy, the US Air Force, the CIA, … . After the USA had detected the missiles, air strike had long been the most popular option within the Executive Committee of senior advisors surrounding president Kennedy (ExCom). However, the US Air Force strongly opposed the air strike because of the uncertainty associated with it. The Air Force could not guarantee that it would succeed in destroying all nuclear missiles at once and the SOP's at its disposal did not allow for the 'surgical' air strike president Kennedy had in mind, only for extensive bombing. On the other hand, the US Navy disposed of a SOP for a blockade and already had considerable strength present in the field.

### 4.6.2.3 Model III: Governmental Politics Model

Allison's third model zeroes in on the actual people that make up states and organizations. The important explanatory concepts include personal power, individual networks, skills of persuasion, charisma and the 'fog of war', referring to people's awareness of their situation to be 'cloudy at best' (Allison and Zelikow 1999, p. 382). Disagreement, miscommunication and misunderstandings are common occurrences. The idea of coherent and transparent state behaviour is totally abandoned in favour of international relations as a messy collage of personal interests, feuds, ambitions, etc.

Implementation of this model for the Cuban Missile Crisis leads to e.g. the following claims. Because of the failure of the Bay of Pigs invasion, Republicans in the U.S. Congress made Cuban policy into a major issue for the upcoming congressional elections later in 1962. Therefore, Kennedy immediately decided on a strong response rather than a diplomatic one. Although a majority of ExCom initially favoured air strikes, those closest to the president—such as his brother and Attorney General, Robert Kennedy, and special counsel Theodore Sorensen—favoured the blockade. At the same time, Kennedy got into arguments with proponents of the air strikes, such as Air Force General Curtis LeMay. After the Bay of Pigs fiasco, Kennedy also distrusted the CIA and its advice. This combination of push and pull led to the implication of a blockade. On the Russian side, Model III reveals Krushchev's huge personal emphasis on Berlin and reports him making a strong link between Berlin and the Cuban missiles.

### 4.6.3  Explanations at the Appropriate Levels of Reality in the Cuban Missile Crisis

The three models offer information at three different levels of reality: nation states as a whole, organisations within these states and individual human actors. Let us see how this information answers the two questions.

#### 4.6.3.1  Why did the USSR put Missiles in Cuba?

Within Model I we can construct several answers to this question. First there is the missile gap explanation: the USSR thought that there was an unfavourable missile gap (i.e. it thought that the USA had more intercontinental ballistic missiles).[8] They placed the missiles in order to close this gap (i.e. to solve this strategic problem). An alternative explanation is that the USSR placed the missiles in order to increase their bargaining power in the second Berlin Crisis (1958–1962). This crisis started in 1958 when the USSR launched a campaign to terminate the Allied presence in West Berlin.[9] In this alternative explanation the strategic problem is "How to get the Allied forces out of Berlin?". One may argue about which explanation is correct (that is an empirical matter) but it is clear that we can answer this question with Model I, i.e. without going to a lower level than the nation states.

#### 4.6.3.2  Missing Camouflage

The second question is a request for an explanation of an unexpected fact. When situations have unexpected outcomes, Model II offers tools to make sense of the puzzlement. As the actions emerging from large organizations can take very strange, unfamiliar forms due to organizational biases and raise serious doubts concerning the rationality of the organizational process as a whole, this second model allows for an account of unexpected facts. More precisely, unexpected events can be explained as the result of the presence of programmes of 'standard operating procedures' (SOP's) which were designed not for the present situation but for some previous circumstance or as outcomes of long and slow processes of organizational struggle bringing about actions nobody might ever have called for. How does this work for the fact that the USSR did not camouflage the construction sites? The implementation of the USSR decision is assigned to organizations that

---

[8] The opposite idea, a missile gap in favour of the USSR was a very common view in the USA in those days.

[9] The 1948–1949 blockade is usually called the first Berlin Crisis. The Berlin Wall was built during the second crisis.

operate by SOP's. As the Soviets never established nuclear missile bases outside of their country at the time, they assigned the tasks to established departments, which in turn followed their own set procedures. The department's procedures were not designed for Cuban but for Soviet conditions. As a consequence, mistakes were made that allowed the US to quite easily learn of the program's existence. Such mistakes included Soviet troops forgetting to camouflage and even decorate their barracks with Red Army stars viewable from above.

This question about the missing camouflage cannot be answered at the level of nation states, because we need organisations and their SOP's. An explanation at the individual level would also be inadequate because the organisational "habits" are important, not the individual actions.

### 4.6.3.3 Appropriate Levels

One of the questions in (EQ3) was: is the scientist considering different levels of explanation? It is clear that for Allison the answer is positive. The example also illustrates that if you take a social phenomenon (like the Cuban Missile Crisis) there are many interesting specific questions one may ask about this phenomenon. Answering all these questions in an efficient way by invoking only one level of reality is usually impossible.

## 4.7 The Efficiency of Propaganda

Our example here is a study by the famous sociologist Robert Merton. During World War II, the American government asked Merton to analyse the success and failure of propaganda campaigns (see Merton 1957, Chap. 14). We use this study to discuss the evaluative questions related to abstraction and level of detail (EQ4). In his analysis, Merton takes individual propaganda documents (e.g. a movie, a pamphlet, a radio speech) as units. Each propaganda document has a specific aim, viz. to convince the reader, viewer or listener to adopt a specific role in the war machine. One of Merton's most interesting examples is a pamphlet that was meant to convince Afro-Americans to volunteer for the army, i.e. to adopt the role of soldier. The pamphlet was a complete failure. It increased the self-confidence of Afro-Americans, but did not convince them to go to the army. So the following question arose:

> Why does this pamphlet increase the self-confidence of its Afro-American readers, rather than convince them to become a soldier?

Merton's method for answering this question (and similar ones) was to divide the document into items. Movies are divided into scenes, while pamphlets contain two kinds of items: text paragraphs and pictures accompanied by captions. The unsuccessful pamphlet contained, according to Merton's analysis, 198 items. Items are then compared with the messages the writers want to communicate to their readers or audience. The aim of the pamphlet was twofold:

(1) to convince the readers that, while Afro-Americans still suffer from discrimination, great progress has been made; and
(2) to convince that these attainments are threatened if the Nazis win the war.

Some of the items were designed to communicate the first message, others to communicate the second message. In general, the items of a propaganda document can be grouped according to the specific message they want to communicate. In Merton's view, a pamphlet will be successful if and only if the items of each group are sufficient for communicating the corresponding message to the audience or readers.

The reason why the pamphlet failed was that the authors tried to communicate the second message mainly by text items. The—mostly lower educated—Afro-Americans did not read the text, they just looked at the pictures. Most of the pictures related to the first theme: they featured Afro-Americans that occupied important positions in the American society. The result was that their self-confidence increased, but they did not conclude from the pamphlet that their attainments were in danger. Hence, they did not volunteer for military service.

Let us now evaluate Merton's explanation. He uses only two kinds of items: text paragraphs and photographs with captions. He makes no further distinctions within text paragraphs (e.g. based on length or topic) or within photographs (e.g. based on size or topic). That was a good decision, given his aim of explaining the failure of the pamphlet: more fine-grained distinctions would lead to the inclusion of irrelevant details in the explanation. The main point in Merton's explanation is that the authors of the pamphlet neglected the fact that the average degree of schooling of Afro-Americans was low. This is their crucial mistake. The classification in two types of items is fine-grained enough to demonstrate the effect of this mistake on the efficiency of the pamphlet. In other words, the answer to the question "Does the theory on which the explanation is based use the right kinds?" (cf. (EQ4)) is positive for Merton.

## 4.8 Proximate Versus Remote Causes[10]

### 4.8.1 The Close-Grain Preference

Several philosophers of social science and scientists defend the thesis that explanations of particular facts should refer to proximate causes (i.e. remote causes are no good at all), or that explanations referring to remote causes are inferior to

---

[10] This discussion is based on Van Bouwel and Weber (2002).

explanations that invoke proximate causes. This point of view is defended by e.g. Jon Elster:

> Both probabilistic models and models involving remote temporal action reflect our ignorance of the local deterministic causality assumed to be operating. (1983, p. 28)

According to Elster, any explanatory factor that is at a temporal remove from the fact explained should be replaced by a factor closer to the fact. This fits in a so-called "fine-grain preference":

> It is the belief that the world is governed by local causality that compels us to search for mechanisms of ever-finer grain (Elster 1983, p. 29)

Two kinds of fine-grain preference can be distinguished: the small-grain preference and close-grain preference. In the social sciences, the small-grain preference advises to look for detailed individualistic micro-foundations that replace macro-level structural accounts. That this is not always a good idea can be seen in Sect. 4.6: in the Cuban Missile Crisis, the appropriate level of explanation depends on the question at hand. The close-grain preference amounts to favouring explanations that provide the detailed mediating mechanisms in causal chains across time. Explanations satisfying this preference will not leave any substantial temporal gaps in the causal chain leading to the event that is explained. In this section we will focus on the close-grain preference and argue that it is a threat of the type discussed in Sect. 3.7.6: it may lead to irrelevant preference rules.

### 4.8.2  An Imaginary Case of Cholera

As we have seen, social scientists and philosophers of social science have taken the position that, *ceteris paribus*, an explanation that invokes the most proximate cause is better than an explanation that invokes a more remote cause. In order to argue against that position, we introduce a fictitious example that is nevertheless realistic (it is based on real causal knowledge). The explanations we compare are structural explanations. Structural explanations must be understood here as *macro-level* explanations that are *causal* (i.e., they explain a social phenomenon by referring to another social phenomenon or structure that caused the first one). This definition puts restrictions on the content of such explanations (e.g., causally relevant information about *individual* agents is excluded, because that is micro-level information), but leaves room for significant differences. For instance, the social event that is invoked as explanans, can be temporally remote from the explanandum, or be very proximate. This is the kind of variation that we need here.

Two neighbouring cities, Koch City and Miasma City, have a history of simultaneous cholera epidemics: every 10 years or so, after excessive rainfall, cholera breaks out in both cities. Suddenly, in the year X, the population of Koch City remains healthy after a summer with lots of rain, while Miasma City is hit by cholera again. Why? Explaining the difference can help Miasma City in the future (therapeutic function). An explanation in terms of proximate causes would look as follows:

(i)   The population of Koch City remained healthy, while Miasma City was hit by cholera, because comma-bacilli were reproduced at large scale in Miasma City, while the number remained very limited in Koch City.[11]

An explanation in terms of remote causes would be:

(ii)  Koch City built a sewage system after the previous outbreak, in the year X-10. Miasma city does not have a sewage system.

Let us look at this example from the perspective of two epistemic interests we introduced in Sect. 3.7.2: improvement and prevention. Explanation (II) refers to a human intervention that was present in one case, absent in the other. Therefore, it can serve as basis for improvement and prevention. Explanation (I) does not have this practical value, because it leaves us with the question of how the reproduction of comma-bacilli can be minimized. This example is not unique: in every case where the proximate causes *do not* relate to human interventions, only explanations in terms of more remote causes are satisfactory if the underlying epistemic interest is prevention or improvement.

In general, we should select the content of our explanation in such a way that it is *adequate* relative to our motivation for asking the question (i.e. relative to our epistemic interests). This strategy sometimes results in explanations containing remote causes (as in our example), sometimes in explanations containing proximate causes. Proximity of causes is not a reason to prefer or dislike explanations: it is an irrelevant property. What we have done here is checked whether irrelevant preference rules are used (cf. the first two questions in cluster (EQ5)).

## 4.9   Where to Find More Examples?

More examples of the pragmatic approach to explanations can be found in articles published by the authors of this book and other researchers at the *Centre for Logic and Philosophy of Science* of Ghent University. Many papers were already mentioned in footnotes in this chapters. All these papers contain more examples than the ones we used here. Many other papers were not used here. People interested in explanation in the social sciences will find more examples in Weber and Van Bouwel (2002), Van Bouwel and Weber (2008a, b) and Van Bouwel et al. (2011). People interested in the biomedical sciences can find additional examples in De Vreese (2008) and De Vreese et al. (2010). For explanation in psychology

---

[11] Note that this is not just a reformulation of the initial question. This answer says that the proximate cause of cholera is bacteria, and not, e.g., the inhalation of bad gases (*miasmata*) as most people believed till the end of the ninenteenth century.

and the cognitive sciences we can recommend Gervais (2013), Gervais and Weber (2013) and Weber and Vanderbeeken (2002). De Winter (2010) deals with explanation in software engineering, Weber and Verhoeven (2002) with explanation in mathematics.

# References

Allison G (1971) Essence of decision: explaining the Cuban missile crisis. Little Brown, Boston

Allison G, Zelikow P (1999) Essence of decision: explaining the Cuban missile crisis, 2nd edn. Longman, New York

Broad W (1985) Star warriors. Simon and Schuster, New York

De Langhe R, Weber E, Van Bouwel J (2007) A pragmatist approach to the plurality of explanations in international relations Theory. In: Proceedings of the 6th Pan-European conference on international relations Conference paper archive of SGIR, Turin.(http://www.sgir.eu/conference-paper-archive/)

De Vreese L (2008) Causal (mis)understanding and the search for scientific explanations: a case study from the history of medicine. Stud Hist Philos Biol Biomed Sci 39:14–24

De Vreese L, Weber E, Van Bouwel J (2010) Explanatory pluralism in the medical sciences: theory and practice. Theor Med Bioeth 31:371–390

De Winter J (2010) Explanations in software engineering: the pragmatic point of view. Mind Mach 20:277–289

Dewey J (1938) Logic: the theory of inquiry. Henry Holt & Co., New York

Elster J (1983) Explaining technical change. A case study in the philosophy of science. Cambridge University Press, Cambridge

Feynman R (1986) Appendix F- Personal Observations of the Reliability of the Shuttle (Reprinted in Feynman 1988). http://history.nasa.gov/rogersrep/v2appf.htm. Accessed 30 August 2012

Feynman R (1988) What do you care what other people think? W. W. Norton, New York and London

Fleissner G, Holtkamp-Rötzler E, Hanzlik M, Winklhofer M, Fleissner G, Petersen N, Wiltschko W (2003) Ultrastructural analysis of a putative magneto receptor in the beak of homing pigeons. J Comp Neurol 458:350–360

Gervais R (forthcoming), Explaining capacities: assessing the explanatory power of models in the cognitive sciences. In: Meheus J, Weber E, Wouters D (eds) Logic, reasoning and rationality. Springer, Dordrecht (in press)

Gervais R, Weber E (2011) The covering law model applied to dynamical cognitive science: a comment on Joel Walmsley. Mind Mach 21:33–39

Gervais R, Weber E (2013) Plausibility versus richness in mechanistic models. Philos Psycho 26:139–152

Gervais R, Weber E (ms.) The Role of DN explanations in biology

Greene N (2012a) Challenger disaster—a NASA tragedy. Part 1: The launch and disaster. In About.Com (Part of the New York Times Company). http://space.about.com/cs/challenger/a/challenger.htm. Accessed 30 August 2012

Greene N (2012b) Space shuttle challenger disaster—a NASA tragedy. Part 2: The space shuttle challenger aftermath. In About.Com (Part of the New York Times Company). http://space.about.com/cs/challenger/a/challenger_2.htm. Accessed 30 August 2012

Haken H, Kelso JAS, Bunz H (1985) A theoretical model of phase transitions in human hand movements. Biol Cybern 51:347–356

Hempel C, Oppenheim P (1948) Studies in the logic of explanation. Philos Sci 15:135–175

Keeton W, Gould J (1986) Biological science, 4th edn. W. W. Norton & Co, New York